Psychology of the Operator of Technical Devices

STUDIES IN SOCIAL SCIENCES, PHILOSOPHY AND HISTORY OF IDEAS

Edited by Bogusław Paź

Advisory Board
Joanna Kurczewska,
Institute of Philosophy and Sociology, Polish Academy of Sciences
Henryk Domański,
Institute of Philosophy and Sociology, Polish Academy of Sciences
Szymon Wróbel,
Faculty of «Artes Liberales» of the University of Warsaw

VOLUME 21

Jan Felicjan Terelak

Psychology of the Operator of Technical Devices

Bibliographic Information published by the Deutsche Nationalbibliothek
The Deutsche Nationalbibliothek lists this publication in the
Deutsche Nationalbibliografie; detailed bibliographic data
is available in the internet at http://dnb.d-nb.de.

Library of Congress Cataloging-in-Publication Data
A CIP catalog record for this book has been applied for at the
Library of Congress.

This publication was financially supported by Cardinal Stefan Wyszyński
University in Warsaw.

ISSN 2196-0151
ISBN 978-3-631-79717-4 (Print)
E-ISBN 978-3-631-80214-4 (E-PDF)
E-ISBN 978-3-631-80215-1 (EPUB)
E-ISBN 978-3-631-80216-8 (MOBI)
DOI 10.3726/b16143

© Peter Lang GmbH
Internationaler Verlag der Wissenschaften
Berlin 2020
All rights reserved.

Peter Lang – Berlin · Bern · Bruxelles · New York ·Oxford · Warszawa · Wien

All parts of this publication are protected by copyright. Any
utilisation outside the strict limits of the copyright law, without
the permission of the publisher, is forbidden and liable to
prosecution. This applies in particular to reproductions,
translations, microfilming, and storage and processing in
electronic retrieval systems.

This publication has been peer reviewed.

www.peterlang.com

I dedicate this book to my siblings: Halina and Krzysztof

Preface

Explaining to the Reader why I wrote this book, I have to get back to an event that took place many years ago, related to scientific cooperation with my friend, an outstanding Polish physiologist Professor Zbigniew Jethon, who in 1976 gave me his book entitled *Działalność operatorowa – nowa postać pracy człowieka [The Operator's Activity – A New Form of Human Work]*. While we were discussing its content, he tried to encourage me to write a similar, complementary one, from a psychological perspective. I was a young experimental psychologist at that time, without experience in applied psychology. It was not until several decades of work at the Military Institute of Aviation Medicine in Warsaw, during which I explored new secrets of psychology and gained knowledge of aeronautical and space technology dynamically developing "right before my eyes", that I was ready to take up this challenge. Moreover, while studying aviation accidents and catastrophes as an expert, I realized how complex and responsible the operator's activity with machines is. Moreover, the longitudinal observation of the professional functioning of aircraft operators in various task situations confirmed my conviction that the most important criterion in empirical sciences concerning the assessment of operators efficiency is not the indeterministic essence of a man, manifested in his intentionality, i.e. in the volitional sphere, but the fact that throughout the whole life of existence (including professional life), a man faces deterministic situations, sometimes being unable to cope with them, thus making mistakes the consequence of which are accidents at work. Looking for my own reason to write this book, I found in philosophy the confirmation that operator's activity is one of the important aspects of human existence, alongside those that have been the subject of philosophical considerations since its inception, such as freedom and its borders, evil, the world and the universe, culture, politics, science, and finally the work itself and its tools. Although my point of view concerns only psychological and correlated ergonomic aspects, a look at this problem from the perspective of modern technical civilization also has something of cultural anthropology, and this may differ from similar monographs of the previous technical "era". What I mean here are the new challenges that the development of the most technologically advanced tools in the history of the Earth's civilization posed to psychology in the 1960s, namely jet aviation and astronautics, which forced the creation of new working tools, thanks to which the modern operator can function and operate efficiently, overcoming certain physiological and psychological barriers. These include "artificial satellites" and "artificial intelligence", which, on the one hand, have pointed to new human limitations in contact with modern work tools, and, on the other hand, to new cognitive possibilities of coping with technical postmodernity. I hope that I will be able to interest the Reader in these new challenges posed to cognitive psychology, some of which are the subject of this monograph. This applies in particular to the references to aviation and space psychology, which may serve as an exemplification of the description of the most advanced tools, much different than the ones from the beginnings of the industrial revolution, which first, at the beginning of the 20th century, overthrew the classic craftsmen responsible for the production of material goods, to replace them with factory workers and engineers, and then, at the end of the industrial era, got replaced by the era of the information society. Distancing myself from the futurological approach, I leave this theme of environmental psychology and the "Concept of Sustainable Development"[1]. I also omit "techno-chauvinism", which was well described by the American Meredith Broussard in the book entitled *Artificial Unintelligence*, published by the *Massachusetts Institute of Technology*, which warns against a new myth (religion) suggesting that advanced technologies can solve all human problems, especially moral problems (e.g., happiness). Another author refers to the latter problems, namely Hannah Fry in her book entitled *Hello Word: How to be Human in the Age of the Machine*. The thesis of this book is that even the most advanced "intelligent machines" do not surpass human intelligence, which also in my opinion is true. These authors rightly point to a certain psychological paradox, according to which in a situation where contemporary man has access to technical tools that allow him to freely transform the surrounding world, he loses at the same time the track of what these tools are intended for, but I leave this thread to the consideration of the Readers after reading this monograph.

Warsaw, 22 March 2019
Jan F. Terelak

---

[1] The concept of sustainable development was a response of the relevant agencies of the United Nations and the European Union to the subsequent civilization challenges that occurred in the second half of the 20th century, threatening the natural environment and contemporary people. Since 1987, following the publication of Our Common Future report, the concept has also taken into account the risks of meeting the needs of future generations.

Contents

1. **The operator–machine relationship as a subject of work psychology** 13
   1.1 Psychology of the operator of technical devices 14
   1.2 Psychology of the operator in Poland 16

2. **Operator–machine system models** 21
   2.1 Technocentric models in industrial psychology 23
      2.1.1 Perceptual aspect of operator's activity 27
   2.2 Anthropocentric models in engineering psychology 34
      2.2.1 Operator-Machine-Environment model (OME) 35
      2.2.2 Operator-Machine-Management model (OMM) 53
      2.2.3 Operator-Machine-Interface model (OMI) 54

3. **Psychological characteristics of the Operator-Machine-Interface (OMI) system on the example of jet plane pilot's activity** 59
   3.1 Concepts of the pilot's role in the OMI system 59
   3.2 Visual attention in operator activity 62
      3.2.1 Role of visual attention in operator's spatial orientation 69
      3.2.2 Oculomotor mechanism in visual attention processes 73
         3.2.2.1 Examples of basic research on the role of eye movements in visual observation 77
         3.2.2.2 Application research on optimization of visual information sources 79
   3.3 Psychomotor mechanisms of operator's action and measurement thereof 85
      3.3.1 Sensomotor reactions and motor coordination 86
      3.3.2 Psychological mechanisms of motor reaction speed 88
      3.3.3 Psychological mechanisms of coordinated sensomotor movements 92

4. Learning new operator activities 97

4.1 Simulation of simple and complex operator tasks 98
    4.1.1 Simulators for learning simple structure sensomotor tasks 100
    4.1.2 Devices simulating operator's physical working conditions 103

4.2 Cognitive models of operator's activity 105
    4.2.1 Conceptual models of operator's activity 107
    4.2.2 Operational models of operator's activity 109

4.3 Training of operator's action in variable working conditions 110
    4.3.1 Training in thermal chambers 110
    4.3.2 Training in a decompression chamber 111
    4.3.3 Training in a high-G centrifuge and a catapult 112
    4.3.4 Training of operator actions in comprehensive simulators as analogues of work experience 114

4.4 Adapting the operator to the real working conditions 121

5. Reliability of the operator in the operator–machine system .. 125

5.1 Reliability as a technical category 125

5.2 Reliability as a psychological category 127

5.3 Psychological concepts of occupational safety 131

5.4 Systematic concepts of occupational safety 135

5.5 Psychological discriminative selection as a method of increasing reliability of the operator-machine system 137
    5.5.1 Analysis of occupations 138
    5.5.2 Analysis of a post 139
    5.5.3 Recruitment and psychological selection 140

6. Computer as a virtual operator serving the optimization of the quality of life 145

6.1 Virtual senses supporting the operator's situational awareness 148
    6.1.1 Virtual electro-optical sensors 148
    6.1.2 Acoustic and radar sensors 149

6.1.3 Biological, chemical, and radiation sensors 150

6.2 Medical applications of a sensory platform 151

6.3 Applications of the sensor platform – Interface in the development of high technologies optimizing the quality of work and life 153

Final thoughts 157

List of figures 161

List of table 163

Literature 165

Index 193

1. The operator–machine relationship as a subject of work psychology

The beginning of the 21st century is an opportunity to summarize the social changes that took place at the turn of the 20th century, taking into account such aspects of social life as the culture of work, leading political systems, the development of technical civilization, etc. Similar comparisons were made at the turn of the 19th century, in which the culture-forming role of the aristocracy as the most educated social class was emphasized. The spatial orientation took into account mainly local centers of power and culture. The disappearance of feudal social relations also in agriculture initiated the creation of a new working class (manufactories, factories), characterizing this period as *pre-modern*. The 20th century can undoubtedly be described as *modern*, providing foundations for and allowing the development of the bourgeois class and mass culture, as well as national capitalism, which, internationalized by Marx's Capital, provided the basis for the description of classical industrial production based on the "work of human hands" operating various machines, the development of which in the second half of the 20th century gave rise to the *postmodern* era, which resulted in a new technical civilization, referring to the intellectual capital of its creators and users, as well as to the scientific foundations of human resources organization and management (Locke, 2003). The postmodern period can be briefly described from a sociological perspective using such attributes as: fragmentation of classical social strata, global orientation of politics and economy, liberal capitalism, information society, modern bureaucracy, growth of mass culture, unlimited possibilities of social communication, fetishization of modern technologies, dictatorship of advertising and fashion, cult of ruthless rivalry, seeking impressions through chemical stimulators (medicines and drugs), excessive consumption of goods, etc. (Simons, Billig, 1994).

In the same period, the humanization of work both in the man–machine system (e.g., engineering psychology and cognitive ergonomics) (Harris, ed., 2001; Sundstrom, 2008) and in the human–organization system (Dunnette, Hough, ed., 1990–1992; Triandis, Dunnette, Hough, ed., 1994) must be noted. Without debating the essence and social consequences of the postmodern era in this book and leaving it to philosophers, sociologists, and political scientists, I refer exclusively to objective scientific achievements in the field of psychology of work and technology, which describe well the functioning of a person in a working situation, achieving his goals with the use of machines and cooperation with other people (Schultz & Schultz, 2008). These objectives include two aspects: individual (satisfaction of personal needs) and social (implementation of social tasks). The relationship between the human being and the goals achieved through the use of technical tools is of interest to the psychology of the operator and its sub-disciplines: Industrial Psychology and Engineering Psychology. From a technocentric point of view, the former dealt with the adjustment of cheap labor force to the operation of machines through qualification procedures called *psychotechnics* (Ackers, 2006). The latter – *Industrial Psychology* together with *ergonomics* – taking the anthropocentric paradigm as its starting point, was aimed at adapting the machine to the psychophysical and mental conditions of the operator through the development of automation (*Cognitive System Engineering - CSE*) and robotization (*Joint Computer System - JSC*) (Sundstrom, 2008). Leaving aside the forecasts for the development of automation and robotization, which are aimed at increasing the attractiveness of work and reducing its inconvenience or even replacing man in his operative activity as an open issue, he draws attention to the fact that this attractiveness of work motivates to choose a specific profession and a place in the organizational structure of this profession. Thus, the attractiveness of the working environment is dealt with by *Person-Environment Theory* (PE) (Holland, 1997), motivation to choose a specific profession by *Person-Job* (PJ) theory (Edwards, 1991), and designation of a place in a specific structure of the organization of this profession by *Person-Organization* (PO) theory (Haslam, 2007).

It is worth recalling a brief outline of the genesis of the psychology of the operator of technical devices and the theoretical and application achievements to date, taking into account the contribution of Polish psychology, referring in detail to the rich review literature of the subject (Chmiel, ed., 2000: Carless, 2005; Kwiatkowski, Duncan, Shimmin, 2006).

1.1 Psychology of the operator of technical devices

It is not easy to establish the beginnings of the psychology of the operator as an important aspect of the classical definition of the psychology of work. Some authors provide contradictory information on this subject. For example, B. Biegeleisen-Żelazowski (1964), the precursor of this field in Poland, points to French sources, which claim that the initiator of the description of man's activity as an operator was the Frenchman Lahy, who since 1908 worked in his first European laboratory for testing the suitability for the profession of railroader, and then a car driver. According to American sources, the world's forerunner was a professor of psychology at Harvard Münstberg University, who in 1912 developed a battery of tests useful for the selection of candidates for tram drivers (after: Caplan, 1987).

Assuming, according to the outstanding Polish physiologist of work Z. Jethon (1976), as valid the thesis that since the appearance of the machine, regardless of its complexity, a *new form of human work has emerged*, requiring specific behaviors, which I call *operator's activity*, I propose to call this branch of psychology the *psychology of the operator*.

Since work plays a very important role in achieving personal and social goals, an important skill of *Homo sapiens* is the use of appropriate technical tools, often very complicated to use. According to the tradition of the psychology of the operator, dating back to the first half of the 20th century, a breakdown into industrial psychology focused mainly on human–machine relations (technical object, physical work environment) and engineering (ergonomic) psychology, dealing primarily with the adaptation of the machine to human capabilities, can be adopted.

The origin of the idea of *industrial psychology* can be traced back to the end of the 18th century, which is connected with the industrial revolution in Great Britain and its socio-economic consequences. The development of large cities stimulating the acquisition of new sources of energy and the use of machinery (e.g., cotton-spinning machines powered first by water and then by steam, or weaving machines, etc.) contributed to the creation, alongside peasants and small craftsmen, of a new social class – contract workers, serving the dynamically developing, especially in the 19th and 20th centuries, industrial production. The British Industrial Revolution, taking place in three phases: providing the driving force, function automation, and workflow control, spreading to other European countries and North America, laid the foundations for a new human activity, namely human *operator's activity* aimed at achieving goals by means of machines. Although at the beginning of its development this field was promoted by American psychologists, the American Psychological Society, which was founded as early as in 1892, did not create an industrial psychology section until the Second World War – although in the 1930s several US universities lectured and trained in industrial and organizational psychology. It was not until the 1980s that the Society of Industrial and Organizational Psychology was established in the USA, with more than two thousand members (Zickar, 2004). Earlier, however, in 1945, along with the dynamic development of aviation technology, the first ever Fitts Human Engineering Division of the Wright-Patterson Air Force Base in Ohio was established in Armstrong[2] Laboratory (Green, Self,

---

2 This section was created on the basis of the Psychology Department of the Aero Medical Laboratory in Wright Field, which functioned until 1945. It was headed by

Ellifritt, 1995), which to this day is the world's leading laboratory in the field of adaptation of aviation technology, characterized by a high level of requirements on the part of the operator to operate it (e.g., the speed of modern aircrafts, exceeding several times the speed of sound, etc.) to human psychophysical capabilities. Meanwhile, in Europe, and especially in Germany and Great Britain, a greater dynamics of industrial psychology development was noted, which was undoubtedly related to the development of the arms industry in these countries after the First World War. In the UK, the National Institute of Industrial Psychology was established as early as 1921, and after the Second World War, the Industrial Productivity Committee was set up, which, stressing the importance of the role of the "human factor" in the work process, was a precursor of research for the humanization of work. However, the Occupational Psychology Section of the British Psychological Association was not established until 1971, which has been and is developing extremely dynamically (Symon, et al., 2006).

1.2 Psychology of the operator in Poland

In Poland, the precursor of industrial psychology was undoubtedly Wojciech Jastrzębowski, who in the middle of the 19th century used the term *ergonomics* for the first time in the world in a treatise entitled *"Rys ergonomii, czyli nauki o pracy"* [*The Trait of the Ergonomics, i.e. Theories about Work*] published in 1857[3] (after: Bańka, 1998). At the beginning of the 20th century, in 1911, Jan W. Dawid published the paper entitled *"Inteligencja, wola i zdolność do pracy"* ["Intelligence, Will and the Ability to Work"], while in 1920 Józefa Joteyko published research on fatigue at work, which were known in Europe, and in 1925 she cooperated in the establishment of the Polish Psychotechnical Society. In the same year, engineer J. Wojciechowski organized Poland's first railway psychotechnical laboratory, and in 1927 the Polish Psychotechnical Society started to publish a

---

captain Paul M. Fitts, professor of the State University of Ohio (Aviation Psychology Laboratory) and the University of Michigan (Human Performance Center).

3 W. Zeidler (2014), who chronologically presents the following sequence of Polish co-creators of operator psychology, is of a different opinion: Władysław Witwicki (1878–1948), Stefan Szuman (1889–1972), Stefan Baley (1885–1952), Józefa Jotejko (1866–1928), Janina Budkiewicz (1896–1982), Bronisław Biegeleisen-Żelazowski (1881–1963), Janina Kączkowska (1895–1978), Estera Markinówna (1903–194?), Helena Słoniewska (1897–1982), Stanisław Studencki (1887–1944?), engineer Jan Wojciechowski (1871–1938), and others.

quarterly "*Psychotechnika*" *[Psychotechnics]* in cooperation with such professors of psychology as: W. Witwicki, S. Błachowski, S. Baley, while at the same time E. Porębski published "*Wykłady z psychotechniki*" *[Lectures on Psychotechnics].* In 1925, the Institute of Psychotechnics was established in Krakow, with branches in Katowice and Sosnowiec, conducting tests for the purposes of selection of railwaymen, tram drivers, car drivers, post office workers, etc. In 1927, a vocational counseling center was established in Lviv to provide services for similar professional[4] groups. A year later, at the Medical Aviation Research Center in Warsaw, Włodzimierz Missiuro (physiologist) and Bohdan Zawadzki (psychologist), the authors of the book *Psychotechnika w lotnictwie [Psychotechnics in Aviation]* (Missiuro & Zawadzki, 1928), organized a psychotechnical unit as a department of the Psychophysiological Laboratory, in which, apart from scientific research, they were involved in the selection of candidates for aviation (Maciejczyk, Terelak, 2017)[5]. Apart from Zawadzki, the second psychologist in this laboratory was Piotr Macewicz, the author of many scientific articles on the *Principles of assessing the ability to work, Methodology of aptitude testing, Mental hygiene,* and others. In 1938–1939 Elżbieta Jaxa-Dębicka, working in the same laboratory, developed a number of psychological tests to assess aviation skills, called "Kanpil" (Polish abbreviation of applicants for pilots) (Maciejczyk,

---

4   A very detailed overview of the development of psychotechnics in Germany and Poland until 2030 was presented by W. Zeidler and L. Helmut (2014). Zapomniany dokument: "Księga Pamiątkowa" Pierwszej Ogólnopolskiej Konferencji Psychotechnicznej, Warszawa, styczeń, 1930 *[Forgotten document, "Memory Book" of 1st Psychotechnical National Conference, Warsaw, January 1930].*
5   The list of all *Psychotechnical Laboratories and Vocational Counseling Offices* in Poland in the 1920s and 1930s includes the following institutions: Institute of Psychotechnics in Warsaw (1919), Municipal Psychological Department in Warsaw (1919), Psychotechnical Department in Warsaw (1925), Institute of Psychotechnics in Cracow (1925), Psychotechnical Tests Office in Warsaw (1925), Psychotechnical Laboratory in Warsaw (1925), Vocational Counseling Office in Lublin (1925), Vocational Counseling Center in Lviv (1927), Railway Psychotechnical Laboratory in Poznań (1927), Psychotechnical Laboratory in Poznań (1927), Psychotechnical Laboratory in Warsaw as an Aeronautical Tests Center (1927), Psychotechnical Laboratory in Warsaw-Praga (1927), Psychotechnical Laboratory in Sosnowiec (1927), Vocational Counseling for Girls in Warsaw (1928), Institute of Vocational Counseling in Katowice (1928), Karol Szymanowski Vocational Counseling Centre in Katowice (1928), Kraushar Vocational Counseling in Warsaw (1928), Counseling for Girls in Vilnius (1929), Vocational Counseling for Girls in Poznań (1929), and Psychotechnical Laboratory at the Railway Directorate in Poznań (1929) (cf. Zeidler, Helmut, 2014).

Terelak, 2017). The development of scientific interests in the field of work psychology can only be observed in the 1960s and 1970s, and it can be associated with philosophical views on work included in the theory of *praxeology*, by T. Kotarbiński, expressed in two treatises entitled *"Traktat o dobrej robocie" [Treaty on Good Work]* (1955) and *"Sprawność i błąd" [Efficiency and Error]* (1960). The first academic textbooks and monographs on work psychology were also published. Moreover, in the 1990s, didactic and research centers (faculties and departments), dealing with widely defined industrial or work psychology, were successively established at all Polish universities. Although in recent years the issue of industrial psychology has attracted less interest in favor of organizational psychology, the psychology of industry and organization is still developing dynamically.

E. Sundstrom (2008) predicts that industrial and organizational psychology are still of great importance in the 21st century for the humanization of work. According to P. Ackers (2006), this is linked to the dynamically developing period of postmodernity, characterized by new technologies and qualitative changes in the urban processes management, which entails a new quality of education for engineers who design modern technical devices in terms of medical and psychological knowledge, especially useful in adapting all kinds of interfaces to human capabilities (Dekker, Rigner, 1999; Nibbelke et al., 1999). This anthropocentric point of view in technical sciences resulted in a renaissance of *engineering psychology* and *cognitive ergonomics* (Harris, 2001). In the theoretical field, we owe this to the departure from the classic "Human-Machine-Interaction model" (HMI), comprising three separate components: man, machine, and their interaction, towards a modern Cognitive Systems Engineering (CSE) model (Hollnagel, 2001). By supporting the operator at the perceptual decision level as well as at the control level (automation), it increases the reliability of the operator's[6] activity (Ruddle, 2001).

Summarizing the short history of the main directions of the development of work psychology and its sub-disciplines, it should be stated that the leading

---

6   It should be noted that already in the 1960s, it was calculated that the simple response time of a pilot flying on an ultra-fast jet plane is too long to avoid a collision with an aircraft flying the same course but from the opposite direction at a supersonic speed of 3 Mach (1 Mach equals the speed of sound, approximately 1200 km/h [760 mph]), if both pilots see each other at a distance of 1 km (so-called psychomotor dead zone). Meanwhile, in 2008, NASA's experimental aircraft X-43A set a speed record of 9.6 Mach or about 10,000 km/h [7,000 mph], improving the previous record from 2004 of 6.8 Mach.

development trend was the employee's empowerment and the humanization of working conditions. Organizational structures bringing together psychologists of work and organizational psychology are also important as a forum for the exchange of theoretical thought and research results. There are two organizational structures in Europe that undoubtedly deserve attention, namely: The European Network of Organizational and Work Psychologists (ENOP), established in 1980, and the European Association of Work and Organizational Psychology (EAWOP). The third international American organization is the International Military Testing Association, which refers to the oldest traditions of the Army General Classification Test (AGCT), and currently brings together the world's best theorists and practitioners in psychometrics (Jones, 2007).

Summarizing the current reflections on the structural genesis of work psychology, it is necessary to conclude this author's own definition, introducing the subject of the book, of the psychology of the operator as a sub-discipline of the work psychology, and at the same time the oldest field of applied psychology. In general, it can be described as a branch of applied psychology that uses the knowledge of theoretical psychology and conclusions from own empirical research to solve multidimensional man–machine relationships, solve all issues of social and humanistic significance in the work environment and in other environments related to work in various ways, e.g., with family, technical civilization, work ethos, economics, etc. The first perspective is related to placing the road transport psychology against the background of operator tasks related to the operation of various types of vehicles: land (rail, road); water (surface, underwater); air (lighter than air and heavier than air); and space (spacecraft, orbital stations). The abovementioned division of vehicles results in the problem of operator psychology, which is defined as an important branch of work psychology, covering both nonspecific issues of the psychology of the operator of machinery and technical equipment, as well as specific problems related to the operation of a specific type of machinery in a specific environmental situation. The psychology of the operator is defined in this monograph as a sub-discipline of work psychology, dealing with detailed issues related to the operation of machines from the perspective of efficiency and safety at work. The nature of the operator's work itself is related, on the one hand, to the complexity of the machine being operated, and, on the other hand, to the degree of difficulty of the operator's tasks and various factors of the working environment. This requires a psychologist to work with a variety of theoretical competences in the field of psychology, as well as very detailed ergonomic and ecological knowledge, acquired practically in the working environment with a specific type of machine and technical equipment. These theoretical and practical competences generally exceed the curriculum of psychological studies. Therefore, it is impossible to be a competent psychologist immediately after graduation, without many years of theoretical specialization (e.g., postgraduate, doctoral studies) and practice, because knowledge and experience concerning, for example, the working environment of an operator working in land, sea, air, and space transport, for example, is so diverse and specific that it requires supplementing knowledge from psychology-related fields, such as: ergonomics, physiology, ecology, law, physics, etc.

The second perspective of approaching the psychology of the operator is related to the model approach to the operator–machine–environment relationship, which, along with the development of human psychology, technical civilization, and ecology, changed its research approaches from a structural to a systemic one. Depending on the adopted perspective, a defined concept of the place and role of the operator in the operator–machine–environment model usually emerges, identifying transactions between them as a new quality, not resulting directly from either the operator's individual features or from the characteristics of the machine or the environment (physical and social)[7] (Anderson et al., 2001; Rogelber & Reeve, 2007).

---

7 The level of complexity of the system approach is well illustrated, for example, by the types of road vehicle operators, for which the appropriate driving license category and medical and psychological examination results are required, reflecting specific skills and technical knowledge: (1) two-wheel motor vehicles (mopeds, motorcycles); (2) motor vehicles (quads, passenger cars, trucks, melexes and golf carts, campers, rally, racing vehicles, etc.); (3) special vehicles (truck tractor, municipal, cisterns, etc.); (4) bus vehicles (coaches, city buses, minibuses, trolleybuses, intercity buses, etc.); (5) low-powered vehicles (agricultural machinery, excavators, tractors, forestry machinery, construction machinery, etc.); (6) special and privileged vehicles (police, sanitary, teaching vehicles, custody buses, military vehicles, etc.).

2. Operator-machine system models

The way of presenting a man in the man–machine relationship requires a precise understanding of the machine, which in the context of interaction is treated as a technical tool of varying degrees of complexity and technological level, used to achieve individual or social goals set by man. The psychological term corresponding to this relationship is the "operator's activity", which was borrowed from a well-known Polish physiologist Z. Jethon (1976), with the only difference that he considers this activity to be a "new form of human work", while I think that it is the oldest form of prehistoric human activity, for which the tool for achieving goals was a machined stone or a stick. The psychology of the operator refers to the oldest evolutionary paradigm of *Homo sapiens* in the human sciences: anthropology, biology, philosophy of comparative psychology, describing the relationship between fitness (physical, biological, and psychological) and its natural determinants of the environment[8]. Some of them are challenges (motivational nature) and other obstacles (stressful nature). In this approach, one can trace the entire history of mankind in the field of operator's activities, the essence of which is the efficient use of both a simple tool, such as a stick or shovel, as well as an extremely complex tool, such as the propulsion of modern jet aircraft or spacecraft. Let us recall that *Homo habilis*, i.e. a skillful man, skillfully using simple tools, preceded in phylogenetic development of *Homo sapiens*, who constructed more and more complicated machines. Therefore, fitness is treated in philosophical anthropology and human biology as one of the most important values in the overall hierarchy of values, as it often determines survival in a threatening situation or development of a technical civilization. Ability as a value is atavistically reflected in ancient and modern civilizations. This has both a subjective and objective dimension. In a subjective sense, people who deviate from the current civilizational model in terms of ability feel handicapped and, consequently, worse than others. In an objective sense, especially in societies with a

---

8 Within the Mediterranean humanistic civilization, the understanding of efficiency comes from Aristotle's definition, which reads: "efficiency is defined as the disposition that is well or badly disposed of in relation to oneself or something else". The fullest distribution of all human skills can be found not in psychological works, but in *Summa Theologiae* by Thomas Aquinas, who divides human skills into intellectual (knowledge, wisdom, conscience, prudence, art) and moral, in other words impulse-sensory (temperance, bravery). (after: Andrzejuk, 2006, pp. 50–51).

Operator–machine system models

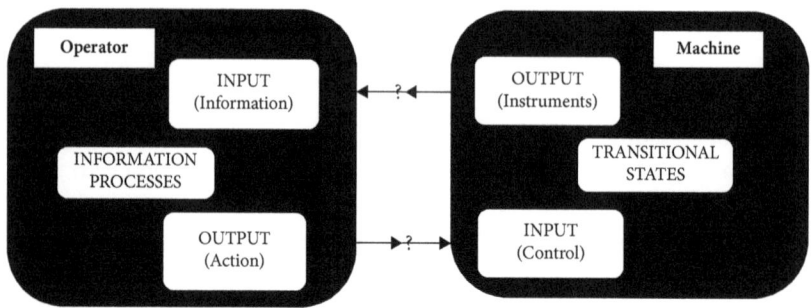

Fig. 1: The basic model of the operator–machine (O–M) (own elaborations))

high level of humanistic culture, fitness – not only physical, but also mental – is a determinant of human value. The divergence from these performance patterns forms the basis for professional selection and development of psychometric tools and simulators for training operator's efficiency. In the psychology of the operator, efficiency is presented from a broad perspective, considering mental fitness and impulse and sensory performance together as *psychomotor performance*. The interest of psychologists in variables responsible for modifying operator's efficiency is reflected in numerous detailed models, which will be the subject of our considerations below. This also applies to the "evolution" of the beliefs of machine builders, which regarded machine as a value in itself (technocentric aspect) and which changed radically under the influence of psychology, giving the machine a servant status in relation to man as a tool to achieve one's goals more efficiently (anthropocentric aspect). Industrial psychology played an important role in this evolution of thinking. Although the general psychological models of the man–machine relationship, regardless of the complexity of the machine, are of a universal nature (Chapanis, 1961), the state of knowledge about man and technology is still changing, generating more and more specific concepts that explain the detailed man–machine relationship (Green, Self, Ellifritt, 1995). One of the most general models of the man–machine relationship is the classic model M–M (*Man–Machine*) or H–M (*Human–Machine*)[9], which is illustrated by Fig. 1.

According to this model, it operates on the principle of behavioral "black box", which describes the functioning of the operator only at the "Input" and "Output", without intermediate processes. There are also no references in this model to the

---

9   In this monograph, we are going to use the original name of these models, namely the Operator–Machine (O–M) Model.

external space (physical, social, and civilizational) in which this system functions and remains in specific relations. Such isolation of the psychological context of work from the technical paradigm of the mathematical communication theory by C.E. Shannon and W. Weaver (1969) is now historical only. Although there are studies that say that the reliability of one component of this system is a guarantee of reliability of the whole system, apart from such a very general formulation, there are no empirical arguments. This is due not only to the oversimplified O–M model, but above all to the difficulty of its detailed and empirical verification, especially in the psychological part of this system. Subsequent theoretical models found in the literature of the subject, although more and more detailed, seem to confirm this pessimism, because they do not go beyond the holistic paradigm, difficult to verify empirically, of the so-called associated work factors, of which there is practically *n + 1*.

2.1 Technocentric models in industrial psychology

Human adaptation to work is a derivative of the more general theory of[10] human adaptation to changing environmental conditions and is the domain of such sciences as: humanities (psychology, pedagogy, sociology, philosophy), natural sciences (biology, medicine, chemistry), and technical sciences. The basic classical paradigm of human adaptation to work in the O–M model was psychotechnics. The original meaning of the German term "Psychotechnics" from the beginning of the 20th century was taken from the views of William Stern (1871–1938), who in 1903 defined *Psychotechnics* very generally as the science of "dealing with people", leading to the optimization of relations in the system: tool – goal, juxtaposing it with the so-called Psychognostics, which deals with psychological cognition. Hugo Münsterberg[11]is considered to be the doyen of

---

10 The terms "adaptation" and "adjustment" are quite often used interchangeably in the literature of the subject. In this book, in the part concerning industrial and engineering psychology, we will use the term "adaptation" of man to machine and machine to man, and in the part concerning the social behavior of the operator, we will use the term "adjustment" according to the canons of social psychology and sociology.

11 Hugo Münsterberg (1863–1916), born in Poland, had a complete knowledge of man, because he studied medicine, philosophy, and psychology in Geneva and Leipzig (in 1885 he obtained habilitation [a post-PhD degree in Poland, Germany, and other European countries; similar to Doctor of Science (DSc) in UK] in philosophy in Leipzig, while in 1891 he obtained habilitation in Freiburg/Br., was nominated as a professor in the same year). As a student and later collaborator of Wilhelm Wundt in Leipzig, in 1888 he founded a private laboratory for experimental research in Freiburg/

psychotechnics.[12] Leaving aside the discussion on the division of psychotechnics into practical, academic, social, or industrial, in this paper we refer to the one proposed in 1930 by the Polish psychologist S. Baley, who defines industrial psychotechnics as a system whose aim is *"to prepare a person to work in accordance with the characteristics of their psychophysical organization, as well as adapting tools and working conditions to that organization, as well as adapting the tools and working conditions to that organization"*. (original citation: *wdrażanie człowieka do pracy zgodne z właściwościami jego psychofizycznej organizacji, jak również dostosowanie do tejże organizacji narzędzi i warunków pracy*, as cited in: Zeidler, Helmut, 2014). We should also agree with Baley's postulate that the psychotechnician should not concentrate solely on working conditions and elimination of people who cannot cope with them, but on changing these conditions. This postulate was fulfilled by psychologists in the so-called era of humanization of work.

The primary objective of this oldest field of applied psychology was to create a battery of tests to assess the potential ability and efficiency of candidates for various operator's job positions. To illustrate this, I cite, following Münsterberg, examples of psychotechnical tests carried out on young German workers (students of the mechanical industry) from the beginning of the last century. The battery, consisting of six tests, included the following capacities and skills of an operator: (1) sensory dexterity (mainly vision and hearing), (2) dexterity (speed and accuracy of movements), (3) attention (ability to observe), (4) technical orientation (understanding instructions and combination of tasks), (5) spatial imagination (mapping of spatial projections of figures and geometric shapes),

---

Br and a year later was appointed by William James to the position of docent at Harvard University. Here, too, he created a laboratory modeled on that of the Leipzig Institute. In 1892, he participated in the congress, where he witnessed the establishment of *American Psychological Association* (five years later he was appointed the President of this association). After returning to Berlin in 1910, he devoted himself to research and didactic work (cf. e.g. Münsterberg, 1914) (after: Zeidler 2011).

12 The term psychotechnics" has been adopted in almost all countries except for the United States of America, where it was replaced by Applied psychology. In Europe, the term "psychotechnics" was preserved until the 1960s, when the name "International Association of Psychotechnics" was changed to "International Association of Applied Psychology", and industrial psychology was given the status "psychodiagnostics with the use of testing instruments" (apparatus for performance tests), which, of course, was not in line with the classical views of Stern and Münsterberg (after: Zeidler & Helmut, 2014, pp. 80–81).

and (6) technical intelligence (general intelligence and memory of spatial forms). This battery included both "pencil & paper" and apparatus tests.

The composition of each battery of psychotechnical tests was adjusted to the work activities required by the machine, which were determined on the basis of the *analysis of the workplace and its course*. The aim of the work analysis was to determine in the first place the share of perceptual (mental) and motor (physical) functions. This required the psychotherapist to conduct *targeted observation* with the use of various techniques and *structured interview*. This also required the technical expertise of the "machine" that the engineer had at his disposal. Therefore, *technical analysis of the work* (issues of the work itself) and *psychological analysis of the work* (issues of human capabilities) have been carried out in a team, which was created by a psychologist and technician (engineer) since the beginning of the history of work psychology. The precursor of the *technical analysis of the work* was F.W. Taylor (1911), who introduced experimental study of working conditions, laying the foundations for a scientific discipline called "work study", developed in Anglo-Saxon countries. The Taylorian study method included, among others, four stages: (1) breakdown of work into functional elements, (2) selection of the best employees for each position and vocational training, (3) creation of conditions of cooperation between workers and managers, and (4) assignment of responsibility for the quality of work to the worker. The precursor of *psychological analysis of work* was a co-creator of psychotechnics, the already mentioned German psychologist H. Münsterberg (1914). This method, in turn, has been the subject of lectures at many American and European universities in the field of work psychology. A psychologist who starts to analyze the workplace must not only spend many days or even weeks together with the operator at the workplace, but above all should have a minimum of technical knowledge before he starts observing. Thus, the observation is aimed in the first place at determining the requirements imposed by the workplace. This is made with the use of lists of requirements for various occupations (e.g., fitter, molder, sheet metal worker, woodworker, etc.) developed beforehand by specialist technicians. The "Set of Professional Requirements" according to R.L. Thorndike and E. Hagen (1959) can serve as an example, including the following requirements for the lathe operator's workstation: (1) physical requirements (muscle strength, fatigue resistance, speed, agility of major muscles, agility of hand muscles, acquisition of new habits); (2) sense efficiency requirements (sharpness of individual senses); (3) perceptiveness requirements (speed of perception for each sense, reaction speed, accuracy of visual differentiation); (4) intellectual requirements (verbal understanding, number manipulation, reasoning, mechanical understanding, spatial imagination); (5) theoretical requirements (pronunciation, knowledge of mathematics); (6) social requirements (favorable external appearance, tact in dealing with people); (7) interests (technical); and (8) emotional requirements (emotional balance). The Polish psychologist T. Tomaszewski (1965) takes a more synthetic approach to the taxonomy of occupations, reducing the entire sociological diversity of occupations to three groups distinguished by: (1) technology of production processes (highly qualified, qualified, semi-qualified, and unskilled workers); (2) required characteristics of the worker (empirical psychograms for various occupations); and (3) activities performed by the worker, including: energy aspect –the dominant type of effort and fatigue –regulatory aspect –perception processes "at the input", decisions and motor activities "at the output" –human–machine interaction aspect –types of communication, complexity of the O–M system, etc.

The aforementioned "founding fathers" of psychotechnics agree that without precise quantitative and qualitative data described for the requirements of each profession and the job position within the profession, it is not possible to determine responsibly the battery of psychotechnical tests, which examine the predispositions for a specific profession. The so-called psychotechnical occupational profiles are a reference point for the results of psychotechnical tests for an individual employee. Obviously it is necessary to examine, with the use of statistical methods, the normal distribution for the whole occupational group and to establish criteria for aptitude assessment. This procedure should be repeated each time for separate professions included in numerous lists drawn up by sociologists in which their comparability is called into question. A separate, long-term, and difficult psychometric procedure is the frequent validation of batteries of psychometric tests in terms of their predictability due to technological changes in machinery (Levine, et al., 1996), an example of which is the dynamic development of aviation[13] technology.

Returning to the O–M model in the analysis of work, it has to be said that psychological knowledge covers only the issues imposed by a technical solution related to the machine design and is limited solely to perceptual (mainly visual) processes related to the control of indicators providing information about the state of the machine and motor activities related to control (Chapanis, 1961).

---

13 For example, in American military aviation, after 50 years of use of battery of pilot IQ tests, they were abandoned as it was considered a poorly prognostic measure (personal information from Edgar M. Johnson, Head of American Military Psychologists and the Director of U.S. Army Research Institute from 2001).

This model determines the range of psychological tests within two areas: (1) perception (senses) and (2) psychomotorics.

2.1.1 Perceptual aspect of operator's activity

In the O–M model, this area of operator's activity involves the senses in the process of receiving information (at the input) and processing it in order to prepare decisions on adequate motor (control) activities. For this purpose, the knowledge of the senses from the oldest psychological discipline, namely *psychophysics* (quantitative parameters of the senses: structure and strength of the stimulus, speed of reaction, modality of the senses, etc.) and *general psychology* (qualitative features of the senses: stimulus as a carrier of specific information), is used (Woodworth, Schlosberg, 1954).

Among the important functions responsible for human orientation in the environment, there are perceptual processes that play the role of mechanisms regulating the man–environment relationship, consisting in the man–environment exchange of information[14]. Contemporary psychology of work, assuming the theoretical basis of the so-called interactionism, claims that human behavior is determined both by a person and by the environment (machine, situation) to the same degree. In an interactive model, this relationship can be expressed by a formula: ***person (personality trait) x situation = behavior*** that can be explained as follows: behavior is a function of the continuous interaction between an individual and situation (machine). An individual is an active party in this set of factors. From the standpoint of the situation, important determinants are those properties that are related to its psychological significance for an individual and for the operation of the machine. From the standpoint of an individual, on the other hand, important determinants of behavior are cognitive processes, in which perception comes to the fore (at the input). At the sensual level, the simplest form of perception consists of sensations and observations[15]. The sensory

---

14 The importance of sensual cognition as a basis for mental cognition was recognized in ancient times by Aristotle in his treatise *De anima*, presenting his view in the famous maxim: "Nihil est in intellectu, quod prius non fuerit in sensu" (There is nothing in the mind that was not first in the sense).

15 *Sensations* can be defined as a simple psychological (cognitive) process that is created by the action of elementary stimuli on a single type of receptor and consists of the reflection of individual characteristics of the stimulus. However, in practice we do not realize the individual sensations but more complex processes called observations. Thus, *observations* are a cognitive process of "reflecting" multimodal stimuli affecting the sensory nervous system.

and observation form of perception, which is the subject of interest in the psychology of work at the level of the H–M model, includes the following aspects of the functional field of the senses: (1) thresholds of senses and (2) variability of stimuli in relation to their location. This first aspect is the subject of research into the oldest psychological discipline, namely psychophysics, which was created at the turn of the 20th century. An important neuropsychological fact then was pointed out that various stimuli from the environment are received through specialized neuron projections (called peripheral neurons), which can be imagined as a cybersystem with one "input" and one "output" of information (sensory). These terms are derived from the technical sciences (communication theory and information theory) and refer to the element of the receptor that changes its functional state under the influence of stimuli. This element is a nerve fiber in each receptor connecting the receptor with the peripheral neuron. The state of the fiber changes under the influence of a stimulus, i.e., the so-called nerve impulse, which is sent to the appropriate area of the brain. The receptor reacts to different stimuli with different frequency of nerve impulses, quality, and strength (intensity). An important property of all receptors is that stimuli of different nature can transpose into a single code common to the entire nervous system (this applies to the frequency of nerve impulses). The specificity of individual receptors has been characterized primarily from the point of view of the informational and energetic properties of the stimulus and receptor reception capacity. This is due to the fact that different types of physical energy (electromagnetic wavelength) are specific to different receptors. Each receptor reacts differently to a specific type of energy depending on the strength of the stimulus and its duration. In general, the duration of a stimulus is the length (time) of a series of nerve impulses sent by the receptor, while the strength (intensity) of the stimulus is the frequency of nerve impulses. Attention is also drawn to the information capacity of the receptor determined by the upper or lower limit of the range of stimulus energy to which the receptors have already reacted or are still reacting. These limits are called either *sensitivity thresholds* or *the lower or upper absolute threshold of the receptor*[16]. The minimum amount of intensity difference causing different reactions is called the *difference threshold* (sensitivity or

---

16 The lower sensitivity threshold is defined as the smallest amount of stimulus energy capable of triggering a receptor response in the form of a specific nerve impulse, and the upper threshold is the largest amount of stimulus energy at which the receptor still reacts adequately to the amount of stimulus energy. Both sensitivity thresholds are different for each type of receptor. The second feature of the receptor is its *sensitivity*, which determines its information capacity. Sensitivity is the minimum size of the

relative). The quantitative determination of the threshold of difference has been known in psychology as *Weber's Law* since 1834 and as *Fechner's Law* since 1860, and is described by means of mathematical formulas, the relationship between the strength of stimuli and the strength of sensations and the ability to distinguish the smallest dieference in the strength of successive stimuli.

Generally, receptors can be divided into two types: exteroreceptors, i.e., receptors located on the surface of the body and receiving stimuli from the environment, and interoceptors, i.e., receptors located inside the body and receiving stimuli from inside the body. Exteroreceptors include *telereceptors* (i.e., receptors perceived from a distance): vision, hearing, smell (fragrance receptors) and *contact receptors* (i.e., receptors receiving stimuli in contact with the receptor surface of the body): taste and skin receptors (touch, pressure, and heat). Introreceptors include *proprioceptors*: balance, joint, muscle, tendon and *viscoreceptors*, located, as functionally specialized nerve endings, in the internal organs of the body, in the walls of blood vessels, in the lungs, intestines, sexual organs, etc. The O–M model focuses primarily on the senses of sight, hearing, and kinesthetic senses, which facilitate interaction with the machine when preparing sensory and motor activities. In this model, constructed on the basis of classical behaviorism in the form of the so-called black box (i.e., without a process mediating between stimulus and reaction), the abstract level of perception including various mental processes such as planning, prediction, evaluation, understanding, and reasoning is omitted. These mental processes are accompanied by *concepts* that are a kind of representation of reality in the human mind, expressed in the O–M model by means of various indicators that indirectly inform the operator about the actual and desired states of the machine. Therefore, the research interests of industrial psychologists during that period focused on functional evaluation of various types of indicators and their visualization. Examples of various forms of instrument's dials with regard to the accuracy of reading the information contained therein are shown in Fig. 2.

As Fig. 2 shows, the most accurate dial, minimizing errors, is the window dial (E), followed by the circular dial (D), the semicircular dial (C), and the linear/horizontal dial; and the linear/vertical dial (A) is the least useful. However, it should be noted that those results included static tests. If we take into account the high dynamics of changes in machine parameters, the window dial is the least useful, and the dial with fully visible scales that allow for anticipation of the

---

difference between two different stimulus intensities necessary to induce two different receptor responses.

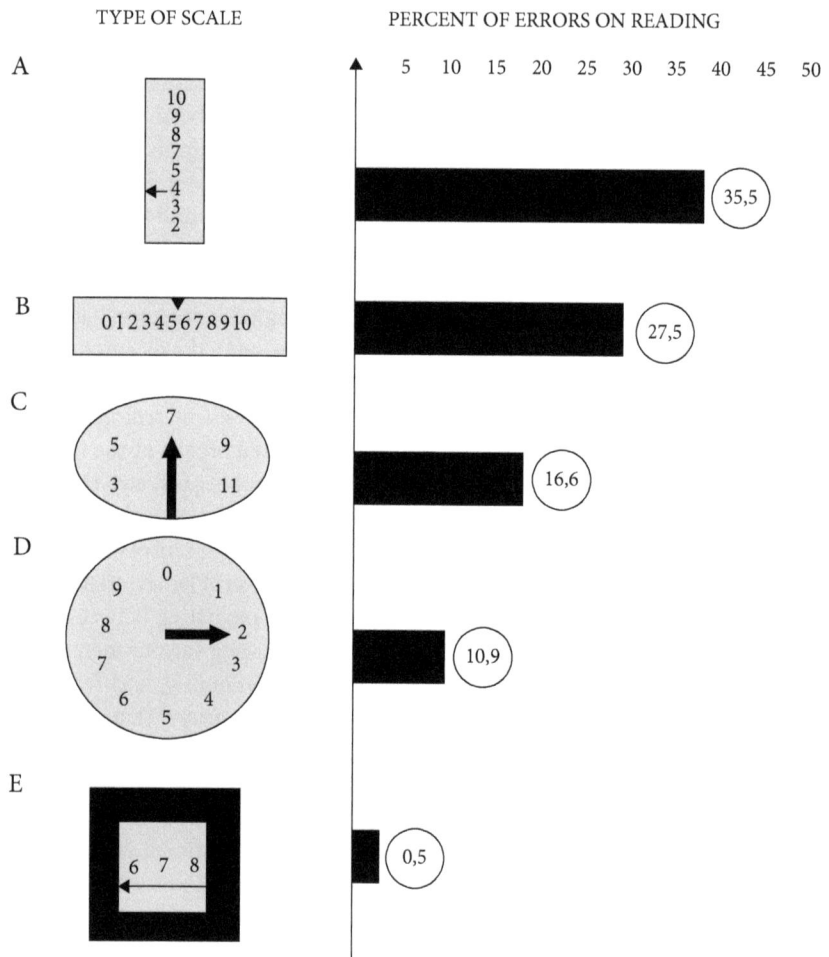

Fig. 2: Accuracy of reading the scale depending on the shape of the instrument's dial (own elaborations)

changing situation comes to the fore. Of course, the size of the dial, which must be a function of the distance from the operator's eyesight in the central field of vision and the type, number, and arrangement of dividers as well as the shape, size, and arrangement of letters and numbers on the scales, is not insignificant. There are two aspects to the correct arrangement of digits. The first concerns the legibility of the digits themselves and depends to a large extent on the direction in which they are arranged in relation to the scale. For example, in the case of dials with movable hands, the digits (letters) should be placed vertically in relation to the whole dial regardless of its shape, and not radially. The latter arrangement of digits is correct in the situation of fixed pointers but moving dials. In addition, digits should always be placed outside the scale lines in relation to the position of the pointer, otherwise they may be obscured by the pointer. The second aspect of the proper arrangement of digits on the dial concerns the sequence of digits of a given scale (so-called scale modules) (e.g., alternating odd and even digits, odd digits only, even digits only). For example, empirical research has shown, among other things, that 10-module scales have the best readability, and among the four types of scales based on this module, the best is the dial in which scale segments (e.g., between 0 and 10) were divided into five component segments (Paluszkiewicz, 1964). The same research shows that the length of the pointer adequate to the dial size is also important, and it cannot be too short or too long. In addition, an important factor for the correct perception of the instrument pointer is the distance between the pointer and the dial background with a scale, and it must be as close as possible. The ideal position of the pointer should be perpendicular to the axis of vision, as it eliminates the illusion of parallax (the impression of a shift of the pointer on the scale compared to the real one). A separate problem is the design of the pendulum *signaling equipment*, as it generally concerns *qualitative* information, i.e. providing global information on a 'normal' or 'threatening' situation (Ely, Thomson, Orlansky, 1956). In the O-M model, two types of pointers were most often used for this purpose: with and without scale. For example, it has been found that a dial without a numerical scale significantly facilitates and accelerates the operator's orientation in an emergency situation. Still in the field of sensory processes, it should be added that we have omitted other modalities of the senses, because in the classical O-M model, based on the theory of information, the emphasis is placed mainly on visual modality. Auditory modality is complementary, and proprioception will be relevant to the discussion of the motor (performance) area of the operator. However, it is worth adding a couple of remarks concerning the determination of quantitative parameters of the visual stimulus duration and intensity (light, sound, etc.), which make the external stimulus a psychological stimulus for the sense of a specific modality. In case of sight, we are informed about this by the Bunsen-Roscoe law: $I \times T = C$, which means that the intensity (I) of light multiplied by its duration (T) determines the stimulus effect and above all its lower threshold of occurrence (Woodworth, Schlosberg 1954). This law applies to all photochemical processes, from the influence of light on the plant growth reaction (phototropism) to the decomposition of the visual purple in the eye, starting the cycle of formation on the retina of the eye. Let us recall that the reaction of purple decomposition in the human eye starts 50–200 ms from the moment the stimulus acts, and this period is called the "critical period" or "functional retinal time". If the duration of visual stimulus is shorter than the critical period, the retinal effect of the stimulus is determined by the simple product of the intensity and duration. The concept of the visual threshold is therefore dynamic and is important in the measurement of sensomotor response time, as the so-called latent time, difficult to identify in classical studies time of simple reaction and choice. It has been empirically verified that with short visual stimulus exposures, when we limit ourselves to small areas of the retina, the *amount of* light plays an important role in determining the thresholds, as described in the so-called Quantum Theory of the Visual Threshold (Woodworth, Schlosberg 1954). There is also in general psychology a concept of *spatial summation of stimuli*, described as *Ricca Law*, which can be written as the following formula: $I \times A = K$ (where: I – brightness of light, A – size of field of view, K – constant factor). This law says that if two surfaces of different sizes have the same brightness, then the eye draws more light from an area larger than the smaller one, and this effect results in the summation of retinal irritation. Thus, the effect of summation of irritation takes place at the retina level (and within the visual thresholds). This law has consequences for the design of illuminated dials with pointers, as well as for the construction of light alarm signals on the machine's hazardous condition.

To sum up the presented design principles of indicators and control devices in the area where the machine is operated, they must be supplemented with conclusions from the psychology of vision, which suggest that devices important for the work process should be placed in the zone of so-called sharp vision, and less important in the area of peripheral vision. An important principle for the arrangement of instrument groups is functionality (relationships between indicator and control devices) and logic (e.g., arrangement of control devices according to function criteria, frequency of exploitation and use, etc.). We will discuss these principles in detail when discussing engineering psychology.

Moving on to the second element of the O–M model, i.e., activities that control the machine, we have to deal with human motility, which includes sensomotor reactions and movement habits of the operator of technical equipment. Activities related to the operation of the machine, involving human motility and psychomotor coordination, are called in classical industrial psychology *working movements*, such as: target, repetitive, continuous, serial, static movements. They were the subject of empirical research of technicians, physiologists, and work psychologists in the field of so-called working grips and coordinated small (precise) movements. In the case of technicians' research, interest was focused on building special models differentiating the working grips in various types of control devices. Empirically many interesting regularities were found concerning the universality of the hand in adapting not only to the correct grip of objects of different shapes (e.g., handwheels, control rods and levers, roller switches, control knobs, cranks and steering wheels, etc.), but also to modulating the force of pressure and to feel resistance from various types of levers controlling the machine. Physiologists of work focused in their interests not only on movements of the joints of the hand or thumb but also on movements of the elbow and lower limbs (e.g., when pressing pedals) from the point of view of anatomy and mechanics of purposeful small and precise movements and brain mechanisms responsible for the speed of sensomotor reaction and coordination of motor activities. The latter problem, in turn, was the subject of research by psychologists of work, who focused on sensomotor reactions and movement habits (Nazarow, 1969). All these three aspects of control activities are complementary in nature, making up the scientific basis for the characterization of human motility from the point of view of their speed (response time); accuracy (quality of work, faultlessness of movements); and strength (effort, fatigue) (Okoń, Paluszkiewicz, 1963). As far as the speed of movements is concerned, psychological studies considered mainly the so-called reaction time, which in these classical studies was defined as a motor reaction (e.g., the movement of pressing a switch), which was a derivative of inaccurate measuring devices (in which the measurement error sometimes exceeded the response time) and the omission of the neuropsychological mechanism of psychomotorics. The latter remark concerns mainly the measurement of the *complex reaction time* (with choice), which is the effect of distinguishing the correct stimulus modality (visual vs. auditory vs. auditory vs. tactile), which is the basis of the correct motor decision, as opposed to the *simple reaction time*, which is the effect of the motor reaction to the single-modality stimulus (either visual, auditory, or tactile). Although in practice we are more often dealing with the reaction time with choice, however, the statement made already at that time by E.J. McCormick (1957) that the time of complex reaction is longer (0.19–0.26 s) compared to the time of simple reaction (0.16–0.21 s) inspired psychologists of work to investigate the neuropsychological mechanism of this state of affairs and individual differences in this area. It was also important information for the designers of such technical devices in which the time limit imposed by the dynamics of the machine requires efficient operator's activity (e.g., development of jet aviation).

To sum up the discussion of the O–M model, we should agree with the opinions that the research work based on this model focuses exclusively on the perception of sensory information and motor activity without any decision element, i.e., that it primarily concerns the search for external relations between selected motor characteristics (e.g., speed, precision) and different working conditions (e.g., shape of the lever, etc.), without detailed reference to explanatory mechanisms. However, the research results obtained at that time were of great practical importance for the designers of technical devices, for a universal description of work activities. However, detailed research on this subject required the use of a more complex model than the classical O–M model, which does not take into account the decision-making processes. This will be the subject of discussion of subsequent models of human operator's activity, taking into account the achievements of engineering psychology.

2.2 Anthropocentric models in engineering psychology

It should be reminded that H. Münsterberg (1914), a proven expert in psychotechnology, strove to make this area more of a *social psychotechnics* nature, because the selection of the right people to perform the required work activities (the so-called subjective psychotechnics) is not appropriate without adapting working conditions and tools to human psychophysical capabilities (the so-called objective psychotechnics).

This view was shared by supporters of engineering and ergonomics psychology. A. Chapanis (1963) made a comprehensive review of world literature on the state of research in engineering psychology until 1963. By pointing out that by 1960 about 3,000 works in the field of engineering psychology had been published and three new titles of scientific journals specializing in this field had been created, including: *Engineering and Industrial Psychology; Journal of Engineering Psychology*; and *Journal of Environmental Science*, Chapanis justified the need for this field of psychology, which is a response to the challenges of increasing dynamics of technological development, especially in military and space aviation in the 1950s.

Engineering psychology, also known as anthropotechnics due to its definition (McCormick, 1964), deals with the adaptation of the machines and technical devices designed to the psychophysical possibilities and psychological needs of man. When presenting the subject and the scope of engineering psychology, the following three approaches, based on different theoretical models, are taken into consideration: (1) narrow, emphasizing the role of psychology in providing knowledge to machine designers, enabling them to adapt the machines to human properties (machine–human model); (2) broad, which goes beyond classical human–machine relations, towards extending research issues towards physical factors constituting working conditions at a specific workstation (human–machine–environment model); and (3) the broadest, including organizational issues related to the selection of various technical facilities for operator workstations as well as the process of vocational training (human–machine–social environment model) (Brannick, Levine, Morgeson, 2007). Since engineering psychology is a response of all human sciences, and not only psychology, to the dynamic development of technology, which preceded some human capabilities to operate it, we assume in this book that it is an integral part of work psychology, whose results of research in the extended O–M model with the so-called intermediary variables (human capabilities) beyond the cognitive value have a great practical significance for designers of machines and all technical objects requiring human service.

2.2.1 Operator–Machine–Environment model (OME)

Issues related to working conditions of operators are primarily the subject of interest of occupational medicine and occupational health and safety (OHS) as well as labor law. Some elements of working conditions are treated as distractors in psychological literature (Howard, Cunningham, Rechnitzer, 1978; Davoren, McCauley, 2007).

Details of the natural and artificial living and working environment are dealt with in a special discipline of psychology, called *sozopsychology* (Bonnes, Secchiaroli, 1995; Avison, 2007), which is most generally interested in the mechanisms of human adaptation to a changing environment. This field of psychology has its own rich literature (cf., e.g., Crawford, 1998). For example, we will briefly discuss the physical factors of the working environment, characteristic of various types of operator's activities, including the conditions of artificial environment (artificial satellites) of aircraft and spacecraft.

(1) Climate factors (humidity, temperature, and wind power). The influence of climate factors on human well-being and efficiency is revealed only when climate zones or seasons change, which trigger adaptation processes called acclimatization. They may also be associated with difficulties in maintaining optimal working conditions in the so-called artificial environment (submarines, planes, spacecraft, and habitats on other planets). For example, temperature has a negative impact on the human body and indirectly on work efficiency when it exceeds the so-called thermal comfort, determined on the basis of subjective evaluation of microclimate conditions of the surrounding environment and objective indicators of work quality. This is the result of the activation of thermoregulatory mechanism, its primary function to maintain a relative balance

between the body's internal temperature and changes in ambient temperature. The condition for the operation of this mechanism to function is a specific temperature level, maintained by the controlling system, the so-called set point of the system (Blatteis, 2007). Climate comfort depends on three factors: ambient temperature, air movement, and relative humidity. An unfavorable system of these three factors may lead to the so-called heat stroke or excessive cooling of the body (Roth, 2007). The maintenance of optimal working conditions in this respect is regulated by appropriate medical standards. Exceeding these standards requires adaptation to new climatic conditions, which is a three-stage process: the stressful nature of climate factors is only revealed by changes in geographical conditions (e.g., during travel related to changes in climate zones) or seasons that trigger adaptation processes called acclimatization (Nelson, Martin II, 2007). Usually three stages are listed in the acclimatization process: the initial one, in which contact with new environmental conditions may cause a deterioration in efficiency; gradual adaptation, the progress of which depends not only on climatic conditions (humidity, air temperature, visible wave radiation, and wind speed), but also on the constitutional properties of man. During this period numerous disadaptive syndromes occur, which may prevent a further stay in a given climate; acclimatization, i.e., relatively constant adaptation to specific climatic conditions. The reverse process is also gradual and consists of two stages: deacclimatization, which takes place after returning to previous climate environment and reacclimatization, which takes place when returning after a longer break to aggravating climate conditions (e.g., the case of workers who worked for a long time outside their basic climate zone and who are characterized by so-called permanent acclimatization in their new workplace).

As has been discussed so far, adaptation to climatic conditions, understood as the process of active adaptation changes, is possible. The acclimatization process itself, understood as a change in metabolic regulation leading to a new level of homeostasis, is, however, highly complex. This claim can be illustrated by the negative nature of the interaction between high ambient temperature and high air humidity and the organism's reaction. The influence of air temperature on the psychophysical state at a relative humidity of 50 % is as follows. Full operational capability is achieved at a temperature of thermal comfort of about +20 Celsius degrees. Slight difficulties at work (annoyance, irritation, difficulty concentrating, reduced mental performance efficiency, etc.) can be observed in the temperature range of +21–26 degrees Celsius. A further increase in temperature to 27–30 degrees Celsius leads to significant discomfort at work (increase in the number of errors, decrease in efficiency of manual work, increase in the number of accidents, etc.). In the +32/-35 degrees Celsius range, there appear physiological disorders (disturbances of water-salt metabolism in the organism, significant strain on the cardiovascular system, strong fatigue, the threat of emaciation), accompanied by a decrease in heavy work efficiency. The highest bearable limit temperature, accompanied by large physiological disturbances, is about +40 degrees Celsius (Radakovic et al., 2007).

Thus, disruptions of subjective thermal comfort are accompanied by changes in physiological functions of the system, associated with the operation of the thermoregulatory mechanism, which depends primarily on the internal temperature of the system and changes in ambient temperature.

The discussion of the effect of low temperatures on human performance at low temperatures (hypothermia) was omitted, as rich review literature is available on this matter (Enander, 1984; Roth, 2007; Taylor, 2007). The human stay in extreme conditions, characterized by low temperature and high air humidity, causes the development of psycho-physiological reactions, leading to the development of relative adaptation to these conditions or disadaptation. The disadaptation indicator is decreased ability to work, and in the case of prolonged body cooling – symptoms of so-called shivering thermogenesis with all its consequences, i.e., disturbance of mental functions, inability to sleep and rest[17]. People on the Earth, thanks to constant wearing of clothing, maintain the temperature under clothes at the level of thermal comfort, corresponding to the temperature of the subtropical climate zone. The need to stay and work in a cold climate (e.g., fishermen fishing in subpolar zones, high mountain and polar expeditions, etc.) justifies conducting experimental research on the process of acclimatization to these conditions. Thus, studies carried out in Antarctic stations indicate, among other things, that during a stay in cold conditions, an adaptation of the human body to cold of hypothermic reaction character occurs (lowering the resting rectal temperature and lowering the temperature curve during thermal load, lowering the thermal threshold of trembling and decreasing cardiovascular functions). Similar studies carried out one year after the return of polar explorers to Poland (at the end of winter) indicate a completely different course of physiological reactions to cold, excluding the hypothermal type of adaption (Kowalski, 1982).

Taking into account the interactions between temperature, humidity, and wind strength, we can distinguish four types of climate on the Earth: two hot

---

17 Exposure of uncovered body of humans currently occurs only among some primitive tribes of Australia (Aborigines) and South America. These people are fully adapted to cold, which is characterized by lowering of the threshold for triggering metabolic reaction (shivers) and lowering of the core body temperature.

(dry, i.e., desert climate, and humid, i.e., tropical climate) and two cold (dry, i.e., land polar climate, and humid, i.e., marine polar climate). Each of them affects the operator's efficiency and well-being if a person is not acclimatized.

(2) Noise. In the working environment, noise is a by-product of the functioning of number of technical devices or human agglomerations. If the auditory stimuli that lead to the perception of sound in the organ of hearing exceed the limits of comfort (subjective and/or objective), they are referred to as noise. Usually noise can be described by two physical parameters: frequency and intensity. Frequency is the number of total acoustic waves reaching the ear within a unit of time and is measured in cycles per second or in Hertz. The intensity, which is the power transferred per unit area, is defined in a conventional logarithmic unit, i.e., in decibels (dB). Distinguishing between harmless and harmful (stress) noises is difficult, because the stressfulness of noise depends not only on the intensity of the sound, but also on the frequency characteristics and duration of the noise as well as the subjective attitude (Cohen, Spacapan, 1984).

Let us recall that the human ear is adapted to the perception of sounds at frequencies of 16–20,000 Hz. Taking into account only the noise level in decibels, it is possible to construct an approximate psychological scale of noise, which is as follows. Noise above 120 dB can be harmful and hazardous to the hearing organs (e.g., eardrum damage). A sound stimulus in the range of 100–120 dB can give painful sensations located in the ear. Noise in the range 80–100 dB has a deafening effect which worsens work performance. Noise in the range of 60–80 dB is perceived as high by other people. The average traffic corresponds to a high noise level of 40–60 dB. Moderate noise has a range of 20–40 dB. Finally, the lowest range of 1–20 dB (e.g., rustling leaves) is called quiet noise (Poulton, 1976). As the above scale shows, it is not possible to clearly determine the harmful nature of noise without the context of the situation (e.g., overly loud music at a disco can be perceived positively in one situation, whereas in another situation as discomfort) for several reasons. However, too high a noise level can affect the performance of the operator by damaging the hearing system (above 120 dB) or causing fatigue to the nervous system (about 100 dB) (Cheveigne, 2001). The stressful nature of noise can be evidenced by the fact that, unlike many other environmental influences, no physiological adaptation to noise occurs. Therefore, the only defense against noise stress is to reduce noise at the source, creating a barrier preventing the spreading of noise, introducing individual hearing protection (e.g., earmuffs, earplugs).

(3) Vibration. In connection with the development of technical civilization and the emergence of a new type of human activity, the so-called operative, associated with the maintenance of technical devices (machines, generators,

road vehicles, airplanes, and spacecrafts), man is exposed to various types of mechanical vibrations transmitted to the human body from these devices. These vibrations with a frequency of 1–1000 Hz are harmful to humans, because they cause the so-called phenomenon of resonance, which occurs at very low frequencies (1–30 Hz) and absorb the mechanical energy of vibrations in tissues (Von Gierke, Mccloskey, Albery, 1991). The criterion for assessing physiological changes in the body under the influence of vibration is primarily the amount of transmitted energy, which can be determined by the following formula: $Q = ISt$, where: Q – amount of energy transferred, I – intensity of vibrations (N/s), S – area of contact (m2), and t – time of exposure.

These vibrations with a frequency of 1–1000 Hz are harmful to humans, because they absorb the mechanical energy of vibrations in tissues, at very low frequencies (1–30 Hz) in particular. Both local vibrations (mainly on hands) and general vibrations have a negative effect on the skeletal-articular system and the vegetative system (ischemic changes associated with vascular contractions, anxiety, the so-called symptoms of Raynaud syndrome). The effect of vibrations, especially vertical vibrations, on visual perception (the so-called blurred vision) and precision of movements in the vibration plane (so-called tremors of legs and hands) were also observed (Turski, 2014). In addition, vibrations induce subjectively perceived symptoms of general discomfort manifested by exhaustion and fatigue (Reinhart, 2008). A technical device operator can be exposed to two types of vibration: *general vibrations – transmitted to the body and head via legs, pelvis, back, or sides from a vibrating surface or seat; local vibrations – transmitted from a vibrating tool to the human body via the upper limbs* (Oborne, 1983).

The only protection against vibrations is their elimination through ergonomic improvement of work stations and tools, which are a source of mechanical vibrations.

(4) Lighting. Poor visibility depends, on the one hand, on the condition of the eye and, on the other hand, on lighting. Poor lighting is treated as a stress factor, because under these conditions people are forced to develop and learn individual strategies to overcome difficulties in receiving visual information. This difficulty is an obstacle and causes errors in work and increases the length of time spent on particular activities. For the above reasons, the lighting required to perform a given type of work has been standardized (Pylyshyn, 1999). It is worth reminding that a lighting standard is essential as the human eye only reacts to a very limited range of electromagnetic radiation frequencies with a wavelength of about 76 to about 38 hundred thousandths of a millimeter. This part of the electromagnetic spectrum is called visible radiation or simply light. Taking into account only the length of electromagnetic waves, it should be remembered that

the eye is most sensitive to radiation of the wavelength located in the middle of the interval mentioned above and that it deteriorates in both directions of this interval, e.g., the wavelength extremes of the visible radiation spectrum correspond to the colors: dark red (76 x10-5 millimeters) and violet (40-38 x10-5 millimeters), while the colors yellow and green are associated with the middle wavelength (Sadowski, 2000). The quality of lighting that determines good vision, especially in artificial work environment, e.g., factory halls, rooms without access to natural light, etc., is also determined by its physical characteristics, which include: luminous flux (lumen – 1 lm), light intensity (lux – 1 lx), luminous intensity of a light source (candela – 1 cd), and luminance (nit, cd/m2). Practically speaking, the eye reacts directly to luminance, therefore the limit of lighting comfort is defined in the literature of the subject by the recommended minimum levels of illumination, which quoting E.C. Poulton (1978) can be characterized as follows. The minimum luminance expressed in nits (cd/m2) for the different visual tasks is: extremely difficult task including the smallest details – 1000-600 (e.g., watchmaker's work); very long task requiring concentration – 600-400 (e.g., fixing stocking); difficult and long task with small details – 300 (e.g., drawing offices); medium difficult task with small details – 200 (e.g., machining); normal task with medium size details – 100 (e.g., work in common offices); roughing tasks with large details – 60 (e.g., assembly of heavy machines); and momentary gazing – 30 (e.g., crossing the corridor).

Night vision minimizes orientation in the environment unless the operator uses a specialized device called a night vision device[18], which allows military and civilian operators controlling technical devices to see in the dark. Night vision devices have been known since the 1930s. For night vision with the use of night vision devices is sufficient to illuminate the field of observation with scattered light coming from stars or the Moon, which after being enhanced several thousand times, allows the perception of shapes without precise details. In absolute darkness, it is necessary to use additional light sources, which are a part of modern night vision devices and generate light invisible to the outside observer (people, animals). Nowadays, there are different types of night vision devices: personal (binoculars, telescopes, and goggles), included in vehicle (cameras) or ships (periscopes), and stationary ones (e.g., cameras on buildings,

---

18 Night vision device was invented in 1926 by J.L. Baird, and after equipping it with an electro-optical converter and image amplifier, it was used in night transport since the 1950s and by the German army on the battlefield since 1941 (cf. Instrument Flying Handbook, 2001).

masts). The quality of night vision comfort is the subject of many detailed ophthalmic and ergonomic studies (cf. Prost, et al., 2005).

The state that causes visual discomfort and drastically reduces the ability to recognize objects is glare. It is the result of unfavorable distribution of luminance or its wide or excessive in space and/or time range. The phenomenon of glare (blinding) and its uncomfortable character are known to drivers, who know that an oncoming car, located more or less opposite, reduces the ability of visual perception. The mechanism of glare consists in the fact that the light that is its source causes reflection on the retina of the eye. The source of intense glare creates a cone of diffused light inside the eye, which we see as dashed lines. A car approaching at night, located opposite us and blinding with the high beams of its headlights, reduces the visibility of the observed image (road, clocks on the car's instrument panel), because the reflection of this light gets into the cone of the diffused light. For these reasons, glare makes it more difficult to observe dark objects than bright objects, as the amount of diffused light, which is the source of glare (cone), is relatively high compared to the amount of light reaching the eye from a dark object (Bruce, Green, & Georgeson, 1996).

(5) Ionizing radiation.

Ionizing radiation is electromagnetic radiation which, in the presence of a radiation source (e.g., a radioactive isotope or an X-ray device), has the ability to cause ionization in the material it passes through, by separating the electron from a neutral atom or a particle outside its area of influence. Thus, ionizing radiation appears as a result of changes in the nucleon system inside a nucleus, accompanied by a change in the energy system. The best known types of ionizing radiation are: (1) Alpha radiation (α) – which is characterized by a high ionizing capability, it has an enormous effect on the human body; (2) Beta radiation (β) – it consists in the emission electrons from the atomic nucleus during the conversion of a neutron into a proton, or positrons during the conversion of a proton into a neutron. The ionization potential of beta particles is less than that of alpha particles; (3) Gamma irradiation (γ) – is connected with the sending by the excited nucleus of an atom during a change of energy state. It is characterized by a high penetration of nuclear radiation, although the ionizing effect itself is small; (4) X-ray radiation (X) – electromagnetic radiation with a short wavelength, it is formed in X-ray tube by bombarding a metal shield (anticathode) with a stream of fast electrons, which are accelerated in an electric field with a corresponding potential difference (order of up to 200 kV) (Poulton, 1978).

The sources of ionizing radiation can be grouped into: natural (occurring in soils, food, plants, and cosmic radiation) and artificial (nonnatural radioactive isotopes, nuclear reactors, X-ray apparatus) (Wolff, 1998). The latter is obtained from nuclear reactions in reactors and accelerators operated by humans (Thierens, et al., 2002). Although they are more and more frequently used in industry, agriculture, and medicine (Di Majo, et al., 2003), since the first nuclear explosion and the recurrent nuclear power plant accidents, such as in Chernobyl (Ivanov, et al. 2004), knowledge of the stressful effects of ionizing radiation is not only common but is also a source of existential anxiety for whole societies.

The biological effects of ionizing radiation on the human body can be divided into two groups: (a) *somatic* (which occur immediately after irradiation of the whole body or are delayed in time: leukemia, malignant bone or skin cancer, cataract, gastrointestinal disorders, infertility) and (b) *genetic* (related to mutations within the genetic material). The magnitude of these changes depends both on the amount of radiation dose and type of radiation, its energy, and duration of exposure, as well as the mass of the irradiated human body and the sensitivity of tissues to radiation.

Occupational exposure to ionizing radiation occurs in: mining of uranium deposits and the separation of radioactive elements, in the production and use of isotopes, in power plants and nuclear-powered vessels, in industrial radiology (e.g., analysis of castings, rolled and welded products, and reinforced concrete structures), in X-ray tube factories, medical equipment, in the analysis of works of art, precious stones or ceramic products, and in clinical radiology (diagnostics and radiotherapy).

The reduction of harmful effects of ionizing radiation on humans is dealt with by radiological protection, setting the permissible exposure doses ([C/kg] – a Coulomb per kilogram; formerly 1R – one roentgen), mainly concerning the evaluation of the degree of ionization of air under the influence of X-radiation or gamma radiation. The biological effects of radiation vary from one radiation type to another. Limit values for ionizing radiation are defined by the International Commission for Radiological Protection and national commissions, which on the basis of permanent research (including the state of the atmosphere) determine the minimum doses of so-called body's threshold doses of irradiation as well as individual groups of tissues and organs (UNSCEAR, 2000). The negative effects of ionizing radiation are accepted, on a lesser evil basis, when fighting cancer cells in the body or taking X-rays (Tubiana, 2003).

(6) Microwave radiation.

In addition to the electromagnetic ionizing radiation discussed earlier, there is another type of radiation transmitting energy at a distance, namely nonionizing radiation, which includes radiowave, microwave, infrared radiation, and visible light. Thus, microwaves are a type of electromagnetic radiation at a frequency of 3×10^{12} Hz, and a length range of $10-4-10-5$ m, which places them between infrared and ultra-short waves, and are classified as radio waves (Poulton, 1978). Despite the widespread use of microwave radiation, e.g., for radio and television communications, radiolocation, etc., biological and especially psychological effects are not yet fully investigated. There are two mechanisms of microwave absorption by matter. The first one is the phenomenon of dielectric losses, mainly due to dipole polarization. The electric field transfers energy to dipoles, which is then dissipated in the material in the form of heat. The second mechanism of microwave absorption, the one used in the production of microwave ovens, is based on ion conductivity, which, when ions begin to move in the material encountered on their way in the direction of the electric field, colliding with other ions (positive move in one direction and negative in the opposite) and with other molecules, release heat energy in the material. Thus, the so-called thermal effect resulting from the conversion of energy generated by electromagnetic fields into thermal energy is well known, the biological effects of which, associated with burns, depend on both the frequency band, intensity, and exposure time. What follows is a degeneration in the cells of parenchymal and myocardial organs, or dystrophic processes in synapses and in the cells of various sections of the central nervous system (Reinhart, 2008). Technical protection measures are divided into three groups of shielding of: radiation source, workstation, and worker (protective clothing). The increasing use of microwave radiation in radiolocation, radionavigation, radiocommunications, and radio astronomy, in addition to the most well-known microwave ovens and lasers in medicine, is already causing less and less discussion about the side effects on the health of the operators of these devices. As we know, the danger results mainly from poor operation of these devices[19].

---

19 Basic applications of microwaves are radar and communication in military and civil applications, on land, water, air, and space (e.g., targeting, missile radio triggers, military electronic reconnaissance, electronic warfare systems, radar speed measurement, collision radar, meteorological radar, geodetic radar, e.g., mapping of the Earth's surface and other celestial bodies), satellite-Earth communications and television, intersatellite communications (with research satellites, space interplanetary probes, and for far-space probe exploration: Viking, Pioneer, Deep Space 1), radio astronomy – by means of radio telescopes, microwave ovens (for defrosting, heating and cooking food), building drying, maser (a device similar to a laser, only that works in the microwave range) as frequency and time standards, electromagnetic weapon (used in the Lockheed F-35 Lightning II multi-purpose aircraft) as a high-energy microwave beam generator, an incapacitating weapon for crowd dissipation (ADS 95 GHz), mobile phones (GSM standard operate in frequencies 870–960 MHz, DCS 1710–1880 MHz

(7) Hypoxia – is a condition caused by a shortage of oxygen in relation to the body's current needs. Oxygen, found in the atmospheric air, is one of the most important physicochemical and biological factors for the proper functioning of the human body, and to be more precise – the maintenance of optimal (from the point of view of the human species) partial oxygen pressure, amounting to 159 mmHg (212 hPa) at sea level. Thanks to homeostatic mechanisms that appeared in phylogenetic development, changes in atmospheric pressure within the range of 800–660 mmHg (1060–880 hPa), characteristic for barric pressure fluctuations on the Earth, do not significantly change the partial oxygen pressure in the organism. However, the problem appears when ascending above sea level with the subsequent reduction in atmospheric pressure and when descending below sea level. Although, thanks to the existence of compensation mechanisms, man can maintain a certain balance within the system, despite changes in the external environment, the scope of it changes in which people can function normally, however, is limited (Molińska, 2015). Taking into account the degree and duration of oxygen deficiency, we can distinguish between acute and chronic forms of hypoxia[20]. Acute hypoxia, the so-called height hypoxia, is characteristic for stays at high altitudes in conditions of reduced atmospheric pressure (e.g., in case of mountaineers or pilots at high altitudes when the cabin becomes depressurized). From the psychological point of view, it is important that the definition of high-altitude hypoxia highlights the phenomenon of functional disorders of a physiological and behavioral nature. Symptoms of **altitude** hypoxia are increasing with the increase in altitude and depend mainly on the length of time spent there and the atmospheric pressure, the ranges of which are as follows: more than 8,000 m (338 hPa) – a critical zone (the so-called death zone); 7,500 m (439 hPa) – a zone of minimum working capacity; above 5,500 m (people do not reside permanently); 5,500 m (540 hPa) – zone of poor working capacity (breathing difficulties, headaches and dizziness, nausea and fatigue); up to 2,000 m (775 hPa) – a zone of working capacity (Fowler, Prilic, Brabant, 1994). For example, the prophylaxis of acute hypoxia in flight conditions above 3,500 m is associated with the introduction of a flight helmet, permanently connected to a

---

and UMTS 2.1 GHz), navigation: Global Positioning System (GPS), wireless computer networks: (WLAN), Bluetooth communication between devices, etc.

20 *Hypoxia*, due to its origin, is divided into the following types: (1) **hypoxenic** (caused by abnormal oxygen distribution in the alveoli), (2) **circulatory** (caused by slower blood flow in the body), (3) **anemic** (caused by low oxygen content in the blood), and (4) **hypotoxic** (related to poisoning of the body, e.g., with potassium cyanide).

tank filled with a helium-oxide mixture, and in the case of mountaineering above 7.5 m a.s.l. – a tank filled with a helium-oxide mixture[21] (Houston et al., 1987).

Knowledge of the physiological and psychological effects of altitude hypoxia is particularly important in situations where a person is staying in an artificial environment, such as in the case of pilots or astronauts, and mainly when technical devices maintaining the optimal range of conditions fail[22].

Hypoxia also occurs in deep-water divers when the gas mixture from the tank, which allows the diver to breathe underwater, is provided in inappropriate proportions, or due to abnormal functioning of the organs, the body will not absorb oxygen properly. Thus, at sea level 0 m (1013.3 hPa), there is work comfort for the majority of the human population, and it gradually deteriorates when descending below sea level: -10 m (2027 hPa) – a zone of working capacity, -20 m (3040 hPa) – zone of limited working capacity; -30 m. (4053 hPa) – earache and rapid fatigue; -40 m. (5067 hPa) – reflex slowdown, visual disturbance; -50 m. and below (6080 hPa and above) – critical zone (sensory disorders, nausea, vomiting, pulmonary edema, respiratory center paralysis, death, and so-called nitrogen narcosis (also called the "raptures of the deep" by divers), causing euphoria and hallucination) (Vaernes, 2007).

Summing up the harmfulness of hypoxia to operator efficiency, attention should be paid to the neurocerebral mechanism, associated with the fact that nerve cells are characterized by low resistance to "oxygen starvation", which first causes hypoxia to disrupt the central nervous system's functions, causing a decrease in the level of perceptual and motor efficiency (Fowler, Taylor, Porlier, 1987). Studies on psychological effects of hypoxia can be divided into the research conducted in natural (mountains) or laboratory conditions (decompression chambers). Observations carried out in natural conditions (mainly of mountaineers) provide data on the occurrence of subjective symptoms of hypoxia depending on its degree and duration of action. The effect of hypoxia associated with staying at an altitude of 3000–6000 m causes drowsiness, mood swings, deterioration of mood, and apathy. At the same time, it is accompanied by a decrease in self-criticism, which in the literature of the subject is described as

---

21 This limited situational awareness of the dangers among mountaineers may come as a surprise, as with their willingness to compete, they practice a form of gambling (reminiscent of "Russian roulette") when climbing the so-called eight-thousanders without an oxygen tank.
22 It is worth remembering that in artificial conditions on board a passenger aircraft the value of atmospheric pressure corresponds to an altitude of 940 m a.s.l.

state of high-altitude euphoria. There are also other depressive reactions. These two ways of reacting to hypoxia can sometimes occur alternately. Particularly dangerous (e.g., in aviation) is the situation of transition from high-altitude euphoria to unconsciousness without any intermediate states. Tests carried out on pilots in pressure simulators show, among other things, that the deterioration of the operator's mental efficiency in acute hypoxia conditions occurs most often at an altitude above 4000 m a.s.l. (Zieliński, Drozdowski & Biernacki, 2014). In addition, it was found in our own studies that the visual attention test is a sensitive psychological test; it consists in subtraction of one unity from one thousand, recording the result. It was found in simulated decompression chamber conditions that this type of "type test" applied at a height of 7,500 is a sensitive tool for early detection of mental impairment, which is indicated by deterioration of the graphical structure of the handwriting and logical errors in a simple arithmetic test (Truszczyński & Terelak, 1996)).

(8) Accelerations. Accelerations occurring at the change of speed or direction of movement during flight cause functioning disorders caused by the inertia force (Von Gierke, Mccloskey, Albery, 1991)[23]. The vector of this force is at the same time the direction of changes taking place in the system. The effect of mechanical forces acting under these conditions on the human body in medicine is defined by the term *load factor*, describing the relationship between the weight of the body at the moment of acceleration and its real weight and the physiological consequences. These consequences depend, among others, on such factors as the amount of the acceleration, its duration, the acceleration rate, and the person's position in relation to the vector of inertia (linear, angular, and centrifugal accelerations) (Mikuliszyn & Żebrowski, 2002). The threshold

---

23 Let us recall that the unit of acceleration is cm/s^2 or m/s^2. However, more often the value of acceleration determined in "g" units, assuming that 1 g is the gravitational acceleration of the Earth amounting to 9.81 m/s^2. Using this unit, acceleration can be defined as a multiple of gravitational acceleration. In accordance with *Newton's second law of motion*, it is possible to calculate the value of the force acting on the body if its weight and acceleration are known. The following formula can be used in these calculations: $F = aG/g$ (where: F – the value of the acting force; a – acceleration in m/s2; G – body weight in kg; g – *standard acceleration due to gravity*). From this formula it is possible, e.g., to calculate the force acting on an aircraft operator performing a flight along a curvilinear track. Assuming that the centripetal acceleration is 50 m/s^2 and the body weight of the pilot is 80 kg, it is possible to calculate that the amount of the centripetal force acting on the pilot is equal to 400 kg, which means that the weight of the body in this situation is about 400 kg, which affects the psychomotor efficiency.

of physiological tolerance of accelerations is their magnitude, beyond which functioning disorders occur, which lead to impairment of physiological and mental fitness in the absence of pathological changes. The limit of biological tolerance is at the same time the line between life and death (Guardier et al., 2008; Wojtkowiak, 2015).

When discussing acceleration from a physiological and behavioral point of view, the direction of acceleration in relation to the x, y, and z axes of the body should be taken into account. From the point of view of aviation medicine, the most harmful are the centripetal accelerations, during which the centrifugal force acts parallel to the long axis of the body in the direction from head to the legs (+x). Under these conditions, changes in the distribution of blood and bodily fluids occur, moving to the lower parts of the body. Thus, there is a drop in blood pressure in the upper areas and their ischemia, while in the area of hips and lower extremities there is an increase in blood pressure, congestion, and stagnation (Wojtkowiak, 2013). For example, in the case of, e.g., 5+Gz, the hydrostatic pressure can be as high as +370 mmHg in the lower limbs, −220 mmHg at the thigh level, −120 mmHg at shoulder level, and 0 mmHg at the head level (Kowalczuk, Puchalska, Palonek..., &, Gaździński, S. P., 2017). The behavioral and psychological effects of acceleration fall into three categories: (a) restrictions on arbitrary movements (related to the increase in body weight) (Truszczyński & Terelak & Jasiński, 2000); (b) *Somatogravic illusions* (related to the disorder of the vestibular system causing the loss of directional orientation in space) (Levkowicz, 2016); and (c) *Loss of consciousness* (caused by acceleration of +Gz). This last disorder has been described in recent years as the "G-LOC Phenomenon" (*abbreviation for: +Gz − Induced Loss of Consciousness*) (Barton, 1988). Hemodynamic disorders occurring at the head level as a result of acceleration in the +Gz direction are symptoms of central nervous system hypoxia. The signal of this type of disturbances is initially the loss of peripheral vision (*gray out*), and a moment later central vision (*blackout*), which is preceded by *tunnel vision* (Terelak, 2002). Visual disorders precede loss of consciousness, which can lead to clinical death in a short period of time (Whinnery & Jones, 1987).

(9) Weightlessness. Experiences with weightlessness are for the human species the newest ones in the whole phylogenetic development, because they are connected with the possibility of overcoming the barrier of the Earth's gravitation. The extremity of this situation is related to such factors as: lack of stimuli registration by appropriate gravitational receptors located in the inner ear, lack of static and dynamic loads of the musculoskeletal system, and elimination of hydrostatic pressure of blood and tissue fluids (Christensen, Talbot, 1986). In the Earth conditions, gravitation force is proportional to the mass of the object

and determines its weight[24]. If the gravitational acceleration value changes, the weight of the body will also change, even though its mass will remain the same. In a *zero gravity* state, an object has mass, but not weight, because the gravity and acceleration applied to, e.g., a spacecraft are even out in orbit (Convertino, 2007).

The medical effects of human weightlessness include, among other things: decrease in circulating fluids, decrease in blood density, bone decalcification, muscle atrophy, noticeably impaired physical ability, etc. This phenomenon is known in the literature of the subject as *orthostatic stress*. Psychological effects are associated with reduced working capacity and subjective discomfort (Kwarecki, Terelak, 1980). For example, the inability of adequate gravitational receptors to register stimuli leads to sensory illusions, spatial disorientation, and response from the vestibular ear (so-called motor disease). The lack of customary load on the musculoskeletal system leads on the one hand to atrophy of the musculoskeletal system, and on the other hand to a significant disorder of visual and motor coordination. The situation of weightlessness as a completely new experience entails the necessity to change not only movement habits, but also existing habits and preferences, such as sleeping, eating, physiological activities, etc. (Suedfeld, et al., 2014). Depending on the length of time a person is in a state of weightlessness, there are two phases: acute stress and relative adaptation. In a stress phase that usually lasts several hours, symptoms of discomfort, anxiety, and spatial orientation disorders (relativity of the concepts of "up–down" or "left–right") appear. Within a few days, the cosmonaut achieves a state of relative adaptation (Hermaszewski, 2013).

A separate problem hindering further exploration of the Cosmos, outside technical and logistical barriers (deVera, Boettger, de la Torre Noetzel,…& Spohn, 2012), are also biological and psychological barriers, related not only to operational efficiency, but above all to being in different gravity conditions on different planets or in conditions of weightlessness during a long interplanetary or astronautical journey to other galaxies[25] (Terelak, 2016; Szocik, Abood,

---

24 It has been assumed that a force of 0.980665 x10 mN (millinewton) exerted at sea level at latitude 45º applies to 1 g. Thus the concepts of mass and weight are different.
25 Other harmful factors of the Earth's environment affecting the efficiency of the operator we omitted are presented here, including: magnetic field, ultrasounds, parasites, bacteria, viruses, molds, fungi, poisons, chemical warfare, etc., which are also important from the perspective of the construction of future space or planetary habitats on the Moon and Mars (Kuznetsova et al., 2017; Pacelli et al., 2019; Goemaere et al., in press). See also the Martian Habitats Project (Cockell et al.,2016) realized by an international team of research centers such as (Cockell et al. 2016): Centre for Astrobiology, School of Physics and Astronomy, University of Edinburgh, Edinburgh, UK.; Austrian Academy

Shelhamer, 2018). A detailed review of the literature on the requirements for the construction of space habitats is carried out by the C.S. Cockell et al. (2016)[26].

10. Toxic hazards

We have omitted a lot of the so-called harmful factors of the working environment, among which toxic hazards are an important group, the hazardousness of which, determined on the basis of experimental and clinical studies, has been known for a long time (Spencer & Schaumburg, 1980). Test results quoted after B.L. Johnson and W.K. Anger (1982) are presented in Tab. 1.

(11) Chronobiological factors of disturbances in work refer to the concept of time, which has an objective (astronomical), physiological, and subjective (social) dimension. These three dimensions are necessary when considering individual phylogenetic and ontogenetic development of a human being. The basic synchronizer of the cyclicality of biological processes, not only for man, is astronomical time, which is detailed in *chronobiology* (Szmigielski, 1974), and *chronopsychology* (Terelak, 1993)[27]. Studies on biological rhythm began with the demonstration of daily body temperature fluctuations in rats and the association

---

of Sciences, Space Research Institute, Graz, Austria; Division of Space Technology, Department of Computer Science, Electrical and Space Engineering, Lulea University of Technology, Kiruna, Sweden; Instituto Andaluz de Ciencias de la Tierra (CSIC-UGR), Armilla, Granada, Spain; Department of Reference Systems and Planetology, Royal Observatory of Belgium, Brussels, Belgium; School of Physics and Astronomy, University of St Andrews, St Andrews, UK; now at the Carl Sagan Institute, Cornell University, Ithaca, NY, USA; Centre for Ocean and Atmospheric Science (COAS), School of Environmental Sciences, University of East Anglia, Norwich, UK; Centro de Astrobiologı́a (CSIC-INTA), Torrejón de Ardoz, Madrid, Spain; Division of Microbial Ecology, Department of Microbiology and Ecosystem Science, Research Network 'Chemistry Meets Microbiology', University of Vienna, A-1090 Vienna, Austria.

26 Cf. Projects: Development of HABIT/ExoMars 2020, co-I Curiosity, co-I of ACE/TGO, co-I of ISEM/ExoMars rover, Analysis of Mars data Astrobiology (habitability, extreme environments).

27 In Polish psychology, the term "chronopsychology" was introduced by J. Terelak (1993), who defines it as a discipline of psychology, remaining in the chronobiological research stream and being one of its components, investigating psychological (behavioral, emotional, and social) consequences of biological rhythms (ultradial, circadial, and infradial) and social synchronizers of human activity, with a view to creating a scientific methodology for evaluation of chronotype (morning vs. evening) and its personality correlates, as well as the cost of psychological adaptation to the conditions of discrepancies between photo-ecological and social factors (e.g., multi-shift work, sudden change of time zones, etc.).

of this observation with changes in "vital activity". The basic synchronizers of functional rhythms in biology are all cyclical changes occurring in the natural environment, e.g., alternation of day and night. From the point of view of the duration of biorhythms, the following rhythms are distinguished: ultradial – i.e., shorter than 20 hours (e.g., peristaltic rhythms, heart rate, respiration, electroencephalographic rhythms); circadial – i.e., around the clock rhythms, lasting from 20 to 28 hours (e.g., sleep – alertness, metabolic processes); and infradial – i.e., lasting over 28 hours (e.g., several days, monthly, seasonal, yearly, several years; menstrual cycle, hormonal cycles, etc.). Rhythms can be determined: (1) endogenously, when they are conditioned by biological oscillations within the organism (e.g., rhythmic activity of nerve cells, organs, and physiological processes) and reflect the natural frequency of spontaneously functioning bio-oscillators; (2) exogenously, when they are conditioned by the variability of external phenomena that do not manifest themselves under constant environmental conditions (Waterhouse, Minors, Waterhouse, 1990).

The most important in the adaptation of living organisms to the conditions of the outside world is the circadian rhythm, in which the social factor is the basic synchronizer in humans, and in animals – the alternation of lighting. The most frequently described biological rhythm is the daily rhythm of body temperature, in which the sinusoidal course of maximum efficiency falls on the afternoon hours and the minimum on the night time (between 3.00 and 5.00 hours) with slight individual deviations. Phylogenetic adaptation of living organisms to the environment took place with the influence of biophysical factors, which are cyclical, such as photo-ecological and thermoecological factors. Because they exert a significant influence on the course of biorhythms, they are called *synchronizers* or *time givers*. Thus the time of the day we do the work of importance as well. This is a particularly serious problem with all types of multi-shift work, the concept that is becoming less and less popular, because during the night period there are many errors in the functioning of an operator as well as work accidents (Akerstedt, Lindbeck, 2007). The problem of daily fluctuations in mental fitness is most noticeable in the context of global statistics of aviation and road accidents at night, where the so-called accident rates are higher than during daytime hours (Buckley, Blanchard, 2007).

The essence of chronobiological stress at work is in the incompatibility of biological rhythms, determined mainly by photo-ecological (lighting), social (time of work and meals), and seasonal factors (seasons and climate zones) (Romero, 2007). This incompatibility leads, on the one hand, to disorders of physiological (health) functions and behavioral (errors in action, fatigue) and emotional (discomfort) effects on the other.

Tab. 1: Health and psychological risks of exposure to certain neurotoxins

Carbon monoxide – CO (Combustion processes)	Reduced attention span;
Inorganic lead (Metallurgy)	Disorientation, loss of vision; nervous disorders of upper and lower limbs;
Inorganic mercury (Production of thermometers)	Hand, face, and leg tremors;
Carbon disulfide (Vulcanization of rubber)	Nervous disorders of hands and feet; psychosis;
Thallium (Glass production)	Nervous disorders in the lower limbic system; nervous disorders of hands and feet;
Organic mercury (Chemical research)	Narrowing of field of vision; nervous disorders of the functioning of hands and feet;
Triethylate (Pharmacy)	General weakness, dizziness;
Methyl bromide (Fumigation)	Nervous disorders of hands and legs;
Carbon tetrachloride (Laundry industry)	Narrowing of field of vision;
Methyl chloride (Manufacture of rubber and plastic products)	Nervous disorders of hands and feet; blurred vision, short-term memory disorders
Trichloroethylene (Degreasing and cleaning of clothing)	Memory and concentration disorders, hand tremors;
Cadmium (Metallurgy)	Loss or weakening of the sense of taste;
Magnesium (Mining)	Psychosis, speech disorders, and hand trembling; decrease in physical fitness, muscle weakness;
Acetone (Cellulose production)	Dizziness; weakness, epilepsy; loss of coordination, blurred vision, abnormal blinking;
N-Hexan (Glue, footwear industry)	Nervous disorders of the lower limbic system;
Toluene	–
Methylbenzene	–
Paints, explosives	Tremors, dizziness, lack of coordination, bizarre behavior;
Aluminum (Mining, processing)	Mental disorders, aphasia, convulsions;
Acetylene tetrachloride (Solvents, repeated use)	Shaking, dizziness;
Styrene (plastics processing)	Loss of short-term memory, nervous disorders of hands and feet;
Cretonne	–

(*continued on next page*)

Tab. 1: Continued

Isobutyl methyl (Centrifugal operations)	Muscle weakness;
Acrylamide (Chemical industry)	Nervous disorders of hands and legs;
Porogat (Pesticides)	Mental disorders, trembling;
Methylene chloride (Pure solvent use)	Illusions, hallucinations;
Pentachlorophenol (In pesticides)	Blind spots, scleroses, and corneal damage; disorders of the autonomic nervous system;
Tetrachlorobiphenyl (Errors in the production of food grade oil)	Nervous disorders of hands and feet;
N-butyl methyl ketone (Varnishing)	Nerve damage to lower limbs;
Polymeric foam	–
Fire extinguishers	Weakened concentration of attention, susceptibility to irritation, blurred vision, eyeball movement disorders;

The factor responsible for interindividual differences in masking circadian rhythms recorded at the level of ability to work is, among others, temperament. One of the first people to draw attention to this was the W.P. Colquhoun (1971). Although individual differences in the daily body temperature rhythm were the first basis for dividing people into "morning" types (these people show high efficiency in the morning and a significant decrease in evening hours) and "evening" types (higher efficiency in the evening) (Blake, 1967)[28]. According to experimental studies, regardless of being a morning type vs. evening type of person, the efficiency at different times of the day varies considerably (Zhanga et al., 2019).

We have omitted another problem, which is important for groups of operators such as transcontinental aircraft pilots and astronauts, is the change of time zones, which entails a reduction in the level of operator's reliability after crossing several time zones in a long-distance flight. This is due to a mismatch between the biological time (measured by the CNS, organs, and cells) and the local astronomical time at the place of arrival and the social synchronizer, which causes the so-called jet-pilot syndrome (*also called: jet-lag desynchronosis, transmeridian desynchronism*, etc.), characterized by symptoms such as insomnia, drowsiness during the day, increased nervous excitability, various degrees of gastrointestinal

---

28 Numerous self-assessment questionnaires are based on the above assumptions. They are designed to identify behavioral syndromes characteristic for the "morning" and "evening" chronotypes (Horne and Östberg, 1976), which correlate with personality traits such as extroversion vs. introversion (Mathews, 1988).

disturbances, inability to perform mental work requiring attention, etc. Many experimental studies confirm the existence of such a syndrome and its sinusoidal structure (Klein, Wegmann & Hunt, 1972; Wegmann, et al., 1983; Samel, Wegmann, 1989). One of the most important factors modifying the ability to work in multi-shift mode is night sleep and sleepiness during the day (Torbjörn, Göran, Mats, 2007). The social and medical importance of desynchronization of circadian rhythms (astronomical, biological, and social) in the context of living comfort and decreased cognitive and psychomotor function is emphasized by the extended health formula proposed by the WHO, which, apart from good physical and mental condition, takes into account the mutual synchronization of crucial biorhythms. The directive also imposes an important restriction on multi-shift work, which is reflected in the ban on night work, as documented in numerous scientific studies (Kwarecki et al., 1982).

2.2.2 Operator–Machine–Management model (OMM)

The *Operator–Machine–Management* (OMM) model is useful at the stage of formulating operator decisions in a situation of communication between people operating in a multi-person cabin (e.g., aircraft). An example of such a model can be the description entitled *Air Force Instruction 36-2243*, under the name *Cockpit/Crew Resource Program*, and *Air Force Instruction AFI 11-290*, in which the components of this model and principles of cooperation of work teams for military aviation were determined (Jederberg, Still, 2002). According to the OMM model, since 1996 the U.S. Air Force has been implementing the *Operational Risk Management* (*ORM*) system, which provides decision-making support between the planning phase and optimization of the operator's executive decisions both in terms of time (speed of decision making) and precision (accuracy of execution), by implementing the following rules: (1) do not accept unnecessary risk; (2) accept risk when benefits outweigh costs; (3) make risky decisions at the appropriate level; and (4) anticipate and manage risk through planning (Karp, Condit, Nullmeyer, 1999).

In civil aviation, in the OMM decision-making model, this third element of the system, i.e., *management*, is defined very broadly as aviation safety management institutions, among which, apart from the classical position of the Air Traffic Controller, operating in the airport area, there are various agencies monitoring the flight and "leading" the aircraft from take-off to landing in the entire airspace (Bourrez, 1999; Stubbs, Danielsson, 2001). Thus, according to the OMM model, the functioning of the operator in at least two planes changes radically. The first level concerns the shift of the center of gravity from psychomotor function to decision making, while the second level significantly changes the personal responsibility of the operator for occupational safety, partly transferring the decision-making powers to the "management center" (Isaac, Ruitenberg, 1999). It also changes the qualifications of personal responsibility for possible "operator error", shifting it towards the so-called human factor, which goes beyond the classic O–M model, expanding it to OMM model. It should be noted that both models are complementary, although each of them has its own limitations, which in the case of the O–M model are associated with personal characteristics, but in the case of the OMM model they concern the characteristics of the organization supporting the decision-making processes of the operator, which, in the network of information flow between particular elements of this organization, causes numerous "misrepresentations" or "delays" in the transmission of information, which makes the so-called human factor in this model more unreliable. Today, due to the use of satellite communication (GPS), operators such as pilots will be able to communicate safely and in real time with all ground-based safety management agencies at the same time, without delay.

The models discussed so far which describe increasingly complex systems of relations between the operator and the technical device focused mainly on explaining the role of particular elements of this system; however, the dynamic development of technical devices, such as jet planes, spacecrafts, nuclear power plants, cars, high-speed trains, etc., largely exceeded the natural psychophysical capabilities of the operator to control them without the support of a computer. This new Operator–Machine–Interface (OMI) model complements the shortcomings of previous models, speeding up the information process on the one hand, and on the other, optimizing the decision-making processes of the operator, especially those working within the time limit (Hawkins, 1993).

2.2.3 Operator–Machine–Interface model (OMI)

The anthropocentric models of the OMI system, as opposed to the technocentric O–M system presented earlier, differ mainly in the replacement of the so-called unreliable human elements or semi-automation or, finally, complete automation of control processes, mainly supporting visual perception through the construction of the so-called synthetic (multi-information) indicators or directive indicators (suggesting the choice of the initial operator's decision) (Parasuraman, Byrne, 2003). The latest OMI models based on a computer module sometimes take over the decision-making and control processes (e.g., modern jet airplanes). The diagram of this intelligent model called *Cognitive System Engineering* (CSE) is shown in Fig. 3.

Anthropocentric models in engineering psychology

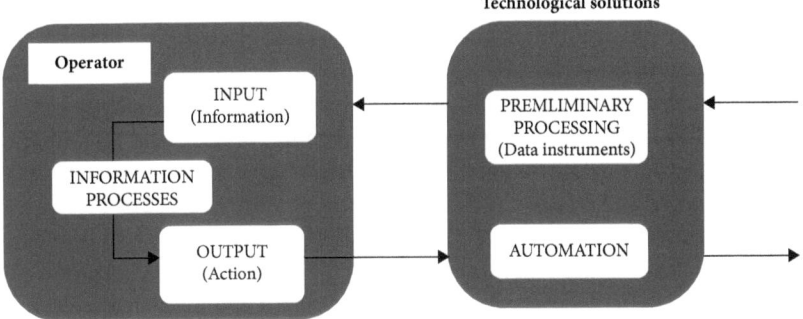

Fig. 3: The intelligent CSE (own elaborations based on Hollnagel (2001))

As shown in Fig. 3, CSE generates technological support from the machine in the form of computer-processed instrument data "at the input" and from automation (semi-automation) "at the output" (Abrams et al., 1991). The development of new information technologies not only streamlines operative activities by replacing classical dials and instruments containing partially processed information (e.g., synthetic dials, grouping on one dial several indicators or directive indicators anticipating variants of operative decisions, etc.) or by relieving the operator of some simple activities (e.g., semi-automation), but also generates new intelligent models integrating functional relations between dynamically changing indicators reflecting the functioning of the machine and the operator's decision-making processes. *Joint System Boundaries* (*JSB*) is an example of such a model, which by taking over perceptual and analytical processes from the operator, as well as control processes partially (automation), moves the operator into a virtual world, where the senses (specific level) are replaced by cognitive processes (abstract level), based on symbolic information about the state of the machine and the external environment (Hollnagel, 2001). This increases the reliability of the operator's activity, especially in time limit conditions, or completely relieves the operator in emergency situations or situations dangerous to health or life (e.g., robotics in the chemical industry) (Ruddle, 2001). The operation of the JSB model will be illustrated on the example of the development of automotive technology, which by placing many sensors in different external parts of the car provides online information about the technical condition of the car and the traffic situation. This changes the classic operator qualifications, optimizing the decisions made and thus reducing errors in action. In addition, the operator has some computer-processed information available and has time to anticipate future machine conditions in relation to the environment, which increases driving safety. In addition, online information on the traffic situation provided by the computer significantly reduces the time needed to formulate a decision, which is a process that includes, among other things: (1) cognitive evaluation of the degree of the task complexity (mental representation of the task), (2) searching for means and methods of task execution (anticipation of conditions determining the results of the action), (3) choosing the means of action (choice of alternative), and (4) evaluation of the decision made, appropriate from the point of view of the desired effects of the action (evaluation of the results). Interface significantly shortens the time to make the final decision because it eliminates many stages of processing a large amount of partial information, which is of great importance in emergency road situations or in the case of piloting a jet plane, flying at a speed exceeding the speed of sound by several times. A computer is very helpful in this respect, providing accurate information, preprocessed "at the input". On the one hand, it increases the reliability of sensory perception (which is very important, e.g., in the case of aeronautical illusions), and on the other hand, it speeds up decision-making processes. In addition, the computer "at the output" often helps the operator in motor activities that require, for example, excessive strength (e.g., power steering or brake assistance) or even relieves the operator in action in the situation of the so-called multimodal task (Cook, Wilson, Proctor, 2001), taking control of an alternative task (e.g., an on-board autopilot taking over control of a military pilot in an air combat situation) (Norman, Draper, 1986). Thus, in comparison to the O–M model in the *Operator-Computer-Interaction* (OCI) system, the mediator between man and machine is another intelligent machine (computer), which "at the input" not only supports the sensory cognition of the changing states of the machine (e.g., power supply of components, rotation, speed, flight altitude, spatial position, etc.) and the changing external conditions of its operation (e.g., the "computer") (e.g., atmospheric conditions, wind force, temperature, pressure, etc.), but also processes operational instrument data into directive indicators suggesting optimal operational decisions regarding the control process (i.e., corrective movements related to maintaining a minimum difference between the flight regimes prescribed at a given moment and the current state of affairs of the machine) (Kulwicki, McDaniel, Guadagna, 1987). In critical situations (e.g., in case of unconsciousness in situations of shock overload +Gz or in case of hypoxia during the loss of cabin pressure), the computer takes over automatic control from the operator. In addition, it is responsible not only for controlling all technical devices, but also for registering the parameters of their operation *online* (the so-called black box) (Barton, 1988).

Anthropocentric models in engineering psychology 57

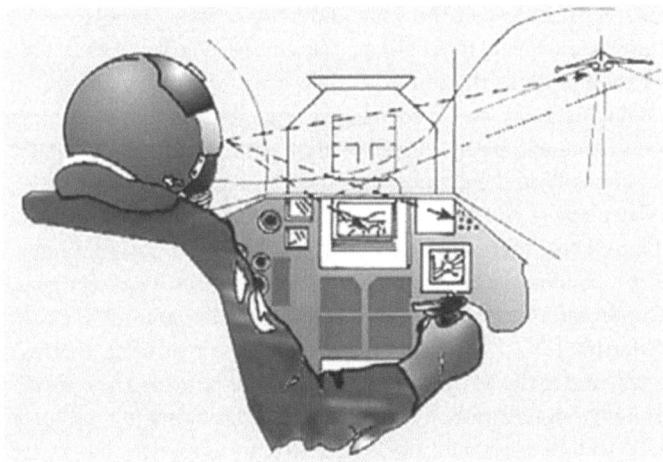

Fig. 4: Virtual Panoramic Display (by Instrument Flying Handbook (2001): courtesy of the Archive of the Military Institute of Aviation Medicine)

The latest generation of instrument information presentation eliminates the traditional instruments that indicate partial information needed for analysis and operative decision making in favor of directive information (preprocessed by the computer) displayed panoramically on the windscreen in, e.g., an aircraft cockpit at the pilot's eye level or even on the windshield of a flight helmet in front of the pilot's eyes (*Virtual Cockpit, Supercockpit, Virtual Panoramic Display – VPD*) (Kocian, 1991). This is a virtual version, which eliminates the pilot's contact with the reality outside of the plane. The most radical versions of virtual instruments practically eliminates the majority of traditional (mechanical, electronic) pilot-navigation instruments, replacing them with the information needed at a given moment of the task to be performed, displayed on the panoramic windshield of the air helmet directly at eye level (Wells, Haas, 1992), as illustrated by Fig. 4.

For pilots or submarine captains, computer graphics provide a reflection of the virtual reality in which the aircraft or ship is currently located, against the background of an initial proposal for the selection of the optimal decision. The "immersion" in the virtual reality displayed on the windshield of the pilot's helmet (*Missiles Streak Across Helmet Visor*) in the panoramic field of view, measuring 120 degrees horizontally and 60 degrees vertically, reflects the state of the plane in a three-dimensional space in the form of color computer graphics. The image changes with the movement of the head as well as compensatory movements with the use of the yoke and pedals. The computer also generates audio prompts, especially in emergency situations (Warren, 1993). For example, steering a submarine in the immeasurable ocean depths or an aircraft in multidimensional airspace (so-called mission mapping) is narrowed down to virtual tunnels (channels, paths, glissades, etc.), which definitely facilitates the very process of controlling the ship or aircraft (Van Paassen, Mulder, 1999; Mulder, Van der Vaart, 2006). At the Armstrong Laboratory at the Wright-Pattersson Air Base in Ohio, USA, there are prototypes of a human brain–computer–airplane interface, which eliminates all pilot-navigation devices for the purpose of controlling the aircraft intentionally, directly by pilot's thoughts (cf. *Biofidelic Virtual Model*) (Martin, 1992). It follows from the above reasoning that in the HCI model, compared to the M–M model, the relations between the operator and the machine change significantly[29], which leads psychologists to reevaluate the psychological selection criteria for position operators of modern integrated systems (e.g., nuclear power plants, controlled production processes, etc.). The emphasis should be shifted from the analysis of sensory performance to decision-making processes and personality disposition (e.g., social maturity, being prone to risk-taking, aggressiveness, etc.), as the role of motor skills is becoming less and less important in the work of operators of complex technical objects, and it is being replaced by the automation of control devices or completely taken over by robots (cf. *The International Journal of Robotics Research*, ed. John M. Hollerbach from the Utah State University). M. Clamann and D.B. Kaber (2004) list ten criteria for evaluating an ergonomic computer-aided solution (*Operator–Computer–Interaction – OCI*): (1) visualization of system's state at any time; (2) matching the system to a realistic description of reality; (3) automatic and arbitrary control of system indicators; (4) consistent standardization; (5) prevention of errors; (6) error prevention rather than error correction; (7) flexibility and efficiency of use; (8) aesthetics and minimization of the control panel; (9) help in quickly identifying, diagnosing, and correcting errors by the user; and (10) *online* assistance and documentation collection.

---

29 The study of increasingly complex relationships between people and computer information is now being addressed by a new interdisciplinary scientific field called *Infonomics*, which is defined, inter alia, as the interdisciplinary study of the social and technological dimensions of the evolution of knowledge in digital society (Heyer, 1997).

3. Psychological characteristics of the operator-machine-interface (OMI) system on the example of jet plane pilot's activity

Before moving to a more detailed characterization of human operator activity, we should start with the identification and conceptualization of the operator in the OMI system.

3.1 Concepts of the pilot's role in the OMI system

As we have tried to demonstrate thus far on the example of various models of the O–M system, it was realized early on that the correctness of the operator's performance depends on the design of the technical devices, the location of the instruments informing about the condition of the machine, the structure of the work activities, and the quality of the operating manual. The conclusions from industrial psychology research presented earlier determined the move away from technocentric models towards anthropocentric models used by engineering psychology and cognitive ergonomics. These disciplines undertook to identify the operator in the O–M system from several interesting perspectives: (1) *Differential perspective* – draws attention to the important fact that the quality of the operator's work, in addition to the excellent quality of the technical object and the level of qualifications, depends to a large extent on individual differences, which are innate to a large extent, such as temperament, responsible for the pace and intensity of work; (2) *Partial perspective* – consists in focusing on certain separate functions and mental processes responsible for the effectiveness of working operations, such as reading the indications of instruments of different shapes, perception of digits and graphical symbols, solving logical tasks, visual-motor coordination, attention distribution, memory, etc. In this case, the analysis of partial aspects of operator activity somehow ignores its place in a specific O–M system. The partial approach still plays an important role in the design of the individual machine elements and is close to the tradition of general psychology and occupational medicine; and (3) *Analogue perspective* – considers each element of operator activity through analogy to a component of a technical device and its mechanical reliability (e.g., the number of operating hours until the first failure of the machine). In this technocentric approach to the O–M relationship, the technical reliability of the machine is to be the standard for the reliability of human activity. With obvious limitations to the analogue perspective, it can be said that certain similarities are only justified when considering highly automated activities of the operator; (4) *Structural perspective* – draws attention first of all to the complex structure of algorithms of operational activities (the so-called checklists) and their relation to the failure states of the machine and self-control within individual work operations. The structural approach, similarly to the partial one, makes it easy to describe externally the very structure of the operator's work activities, but this applies only to technical devices of a specific construction and people with a specific structure of abilities and personality (the paradigm of individual differences); (5) *The system perspective* – in contrast to the approaches discussed above, it is characterized by a holistic analysis of the O–M system. The starting point is not the analysis of elements of the structure and algorithms of the operator's operation, but a *system feature* as a new quality, not resulting directly from either human or machine characteristics, but from interactions between elements of the system. The systemic approach is difficult to quantify because it requires very detailed knowledge of each element of the system, which exceeds the competence of a single scientific discipline, and therefore requires an interdisciplinary approach, which does have certain methodological limitations.

Depending on the adopted research perspective, a specific concept of the place and role of the operator in the O–M model is drawn up. The most common operator concepts are as follows:

(1) *Concept of the operator as the guarantor of failure-free operation of the technical device.* This concept defines the operator as being responsible for the failure of the machine only at the stage of its preparation for operation (e.g., the engineer and technician responsible for the vehicle approved for use under the guarantee).

(2) *Concept of an operator carrying out repairs of technical equipment.* This concept draws attention to the role of the technical staff, who are only responsible for replacing faulty machine components under stationary conditions[30].

(3) *Concept of an operator performing periodic preventive maintenance of technical equipment.* According to this concept, the operator is responsible for the technical condition of the machine under realistic operating conditions.

(4) *Concept of an operator testing technical reliability after repairs and preventive maintenance under realistic operating conditions.* This concept

---

30 This concept has its adversaries, who believe that any human intervention in a technical device, even conducted in goodwill and with high professional competence, does not increase but reduces the reliability of the machine.

draws attention to the responsibility of the operator for the final diagnosis of the technical condition of the machine just prior to putting it into service[31].

(5) *Concept of the operator directly controlling the technical object.* This concept assumes the existence of an anthropotechnical system, which consists in assigning man an active role in the O–M system, and recognizing it, with all its sensory limitations, which are in modern technical devices supported by computer and automation, as a guarantor of reliability of the entire OMI system. This concept is based on the assumption that in unforeseen situations, only an operator with his intelligence and creativity is able to maintain a high level of security of the system, in which there is a need to include one more element of the model – management, described by the next concept.

(6) *The concept of operator as a manager supporting the functioning of the O–M system.* The concept refers to a servant role for the operator directly operating the machine, social support from a specialized manager of an individual or institutional nature (e.g., a co-driver working with a racing driver or air traffic controller, etc.).

Regardless of the definitions and concepts of the operator, three basic types of operational activity can be distinguished in their activity, distinguished according to the essence of the task, the form of exposure to information about the condition of the machine, and the environment as well as the importance of the activity for the operator.

(1) *Cognitive activity* – focused primarily on the processes of visual attention (perception of information and its initial evaluation). Accuracy and speed at this stage depend on the type and form of sources of information about the condition of the machine, on the one hand, and, on the other, on the individual features of the operator (e.g., type of nervous system, aptitude, level of proficiency, etc.). This type of activity of the operator is not yet fully automated, as it is treated in the O–M system as a "vigilance indicator" of the active operator in the entire system. An example would be the need to maintain pilot activity at an appropriate vigilance level during

---

31 This applies to so-called test driving or test flights, during which the machine is operated within its extreme operating range in order to detect "hidden" faults. Therefore, job of an operator of, e.g., a "Formula 1" car (test driver) or a jet plane (test pilot) treated as a high-risk profession.

a transcontinental flight with the use of an autopilot, where the pilot's vigilance is tested by ground handling that sends an auditory signal from the nearest radar beacon, which must be confirmed by pressing the appropriate response key and by voice.

(2) *Corrective activity* – including executive activities, it is characteristic of the so-called tracking of information according to pre-established instructions and correction of discrepancies between the current instrument indications and the preset ones. The correct execution part of this type of activity can take different forms: verbal message, own logical operation, and motor reaction. The latter form – corrective action – is more and more often subject to automation.

(3) *Conceptual activity* – takes into account the transformation and partial decisions developed earlier together with corrective actions in the final conceptual model, which contains a definite idea of the final (deliberate) result of the work of the entire O–M system. It is a completely new quality relationbetween the information model, including machine features and states, and the conceptual model, taking into account the operator's subjective action plan. This subjective thread of the conceptual model makes it impossible to treat this model as a simple reflection of the information model. Otherwise, the operator could be eliminated from the entire "O–M" system. This intentional model of operator activity requires taking into account not only the quantitative features of cognitive processes (perception, memory, attention, and thinking) and psychomotorics, etc.,but also qualitative characteristics related to, among others, motivation, system of values, etc.

As we tried to demonstrate in Chapter 2, the psychological characteristics of the operator activity in each of the several models of relations presented earlier: operator–machine, covered at least two areas of operator activity, namely visual perception at the "input" and at the "output" – psychomotor coordination.

3.2 Visual attention in operator activity

Since visual attention plays a major role in the operator activity "at the entrance", let us look at the complicated psychological mechanism of vision. Let us recall that the visual stimuli for the eye are electromagnetic waves, covering a small part of the spectrum of radiation in general. This part can be extended by using suitable technical devices for night vision and infrared. Under the conditions of a natural operator activity, at the visual perception level, the functional vision system is considered at three levels of visual stimulus analysis. The first level, responsible for *daytime vision* called *photopic vision*, provides comfort of visual perception in daytime photo-ecological conditions, using two subsystems: the *focal vision* and *peripheral vision* subsystem. The *focal vision subsystem* includes a small angular area of sharp vision, and is responsible for the identification and recognition of visual stimuli (answering the "What?" question). The *peripheral vision subsystem* is responsible for determining the position of the visual stimulus in space and for general visual orientation (answering the "Where?" question). This subsystem is based on the analysis of relatively large stimuli, omitting small details (e.g., observation of the roadside with the so-called corner of the eye). It operates on an all-or-nothing basis, and therefore the participation of conscious processes in the activities of this subsystem plays a minor role, if any. This subsystem is not very thoroughly examined experimentally compared to the central vision subsystem, which plays a key role in the examination of visual attention. The second level, called *mesoptic vision*, is responsible for the relative comfort of vision at dusk, while the third level, called *rod vision*, is responsible for the discomfort in visual perception at night. Examples of night vision comfort without night vision support can be described on a scale from -5 to +5, as follows: (-5) snow visibility against cloudy sky, (-4) snow perception in starlight, (-2) snow in moonlight, (-1) snow in twilight, (0) a sheet of paper at a 30 cm distance from a candle, (+1) a sheet of paper held 120 cm from a 100W bulb, and (+2) a sheet of paper visible in the sunlight (Jacob, Jeannerod, 2003). As the above suggests, there is a low visual efficiency under typical night vision conditions, which is related to optical phenomena such as: (a) disturbance of depth perception (which is dangerous, e.g., in the perception of the runway of a pilot approaching landing, or which causes many visual illusions); (b) reduction of the ability to perceive motion, especially for long distances (e.g., in daytime conditions, the angular velocity necessary for registering motion is 0.5–1 degree/sec, and in night conditions it must be 10–20 times higher); (c) the blinding effect (glare); (d) slow adaptation to darkness (e.g., only after 30 min does it reach 80 % of the possible adaptation under these conditions); (e) night-time shortsightedness; (f) lowering of the sharpness of vision; (g) narrowing of the field of vision; and (h) central scotoma. This low visual efficiency under typical night vision conditions as well as the glare from light sources and the resulting blindness were one of the main reasons for the development of night vision assistance devices that transfer the night vision range into dusk vision range by means of night vision and thermal vision devices, which are useful in conditions of extreme smoke or fog density and darkness (used by, e.g., the army, emergency services, etc.)[32]. Although the image perceived by means of night vision devices does not reflect the full comfort of daytime vision, it makes it much easier for the operator to distinguish many details in the field of vision, although the image is slightly different from the real one.

In order to illustrate the evaluation of visual performance in various conditions of visibility during day, dusk, and night, it is worth quoting the model of experimental research on the evaluation of vision efficiency wearing goggles (Prost et al., 2005; Kuliński et al., 2005; Stasiak, 2009). The research was carried out using the HIPERION flight simulator, which enables the presentation of the synchronized image simultaneously on five collimated screens in the range of 180° horizontally and 30° vertically. A system of mirrors gives the illusion of perspective and an infinite visual distance. The flight of a combat aircraft was simulated according to the *FlightGear* program, which allows for a faithful representation of the details of the Earth's terrain appearing on the screens. To improve night vision, PNL-3 goggles mounted on a pilot helmet were used. The aim of the study was to answer the following research questions: (1) How sharp is the vision and the sense of visual contrast in static and dynamic conditions (on a HIPERION flight simulator) in day, night and night vision with night vision support? (2) Does the additional visual task (the Landolt "cube") affect the operator efficiency of a pilot flying on a HIPERION simulator? The quantitative evaluation covered: (1) static sharpness of vision; (2) sharpness of vision during flight on a simulator (measurement is carried out using Landolt optotypes, so-called cubes seen from different distances under different angular size); (3) sense of contrast during flight on the simulator; and (4) influence of an additional ophthalmic task (distractor) on the quality of performing an aviation task. The subjects were pilots with proper, healthy eyesight (according to the current regulations for night vision goggle flights). The studies were carried out in three different photo-ecological conditions: (1) *in daylight vision conditions* in a flight helmet, with ambient lighting of approximately 100–200 lux (behind the screens) and a luminous intensity of approximately 70 candelas; (2) in *nighlight vision conditions* in a flight helmet, with lighting of 0.01–5 lux (behind the screens) and luminous intensity of the screens about 20 candelas; (3) in *nighlight vision conditions*

---

32 The most typical night vision devices belonging to the U.S. Army and Air Force include: thermal vision devices; night vision goggles (NVG); and devices composed of thermal, radiolocation, and laser systems (e.g., LANTRIN, TRAM, PATHFINDER, ATLANTIC) (according to Instrument Flying Handbook, 2001).

with the use of night vision goggles (NVG) with ambient lighting about 0.01–5 lux (behind the screens) and luminous intensity of the screens of about 20 candelas. (2) nonassisted night vision (the test procedure is similar to daylight conditions, but before the test, the subject is adapted to darkness for 25 minutes); (3) night vision with the use of night vision goggles. Operations in these three photo-ecological conditions included simulated helicopter flight along the marked route following the orientation signs appearing in the field of view, which point the direction in which the pilot should fly. At the same time, Landoldt optotypes are displayed on the cube from the beginning of the flight. The subject's task was to read out the position of the gap in the Landoldt ring as quickly as possible. The pilot is to press an appropriate button on the joystick the moment the gap is recognized. If correctly recognized, the "cube" changes its position by 90°. The next optotype with reduced contrast will be displayed on its next side. If the reading is incorrect, the subject will switch off the signaled recognition and the unchanged sign will move further towards the aircraft. Only when correctly recognized, the "cube" will change its position. The next "cube" will appear only after the disappearance of the currently read cube. The following aspects are evaluated: (1) metric value of the size of the recognized optotype, (2) response speed time with correct recognition of optotypes, (3) size and percentage value of contrast at which optotypes were read, and (4) response speed time with correct recognition of optotypes.

The study confirmed the deterioration of vision at the level of the examined parameters in night vision conditions, both with and without night vision goggles. However, the use of night vision goggles significantly improves the quality of vision compared to night vision without the use of goggles. In addition, it was found that an additional visual task worsens the operator efficiency of the pilot. This is in line with the literature that explains the mechanism of functioning of the operator based on three models, referred to as *dual/multi-task interference* (Wickens, 1980; Pashler, 1994; Kinsbourne, 1981). The diagram of these models is presented in Fig. 5.

As Fig. 5 indicates the presented models are based on the assumption that, i.e.,: (1) *Capacity sharing* – when performing more than one task at the same time, the potential for data processing and responsiveness is shared between these tasks; (2) *Cross-talk* – simultaneous processing can only involve the same meaningful signals, otherwise the processing is significantly degraded; (3) *Bottleneck* – there are situations where parallel processing cannot take place, because it leads to a deterioration in the performance of one of several tasks, and in extreme cases – all tasks. The first two theories make the ability to perform several tasks simultaneously dependent on the efficiency of hypothetical mental

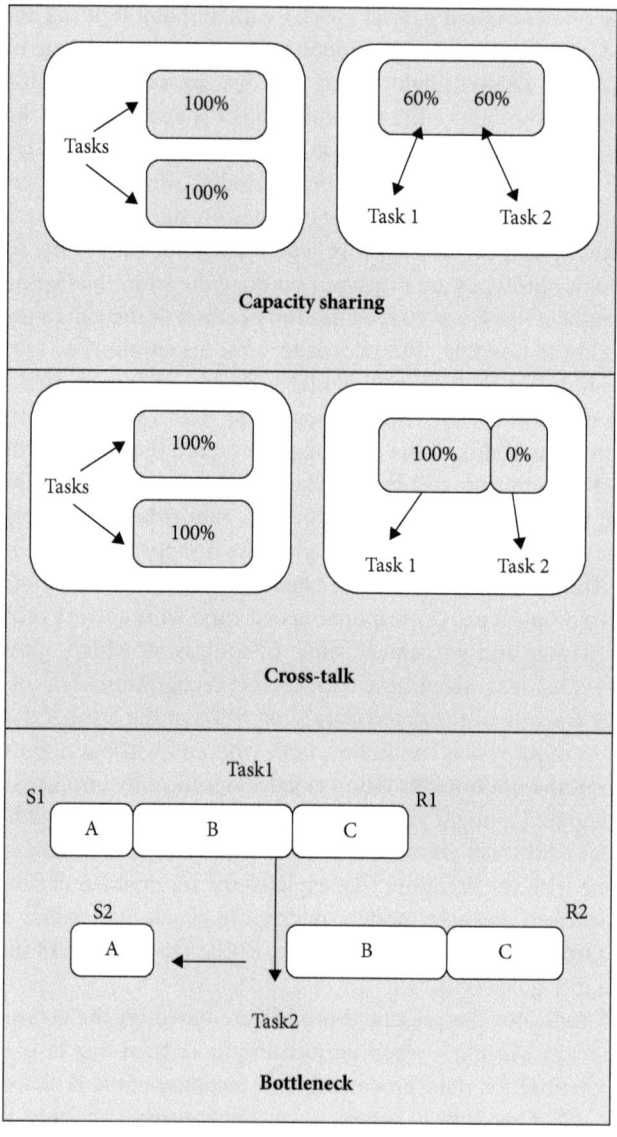

Fig. 5: Models for performing two simultaneous tasks (own calculations based on: Szczechura, Malawski (1999))

mechanisms, while the third focuses primarily on the content of the processed information (Kinsbourne 1981, Navon and Miller 1987).

Verification of the presented models in conditions similar to real conditions was the subject of research carried out by J. Szczechura and M. Malawski (1999), which was conducted on 96 people aged 19–22 (candidates for an aviation school). The subjects' task was to perform a cycle of five flights on the "Iryda - JAPETUS" flight simulator of a training and combat plane I-22 in accordance with the previously provided flight schedule (flight time was about 30–40 minutes). The visualization of the flight was displayed on collimated screens, which reflected the actual flight conditions faithfully. Two situations were tested: (a) an additional visual task (a sequence of digits displayed on the screen in the upper right corner) and (b) an additional auditory task (evaluation of three-digit sequences transmitted in headphones stochastically at intervals of 2 or 4 seconds). The pilot's task was to press the button on the joystick as soon as possible if the sequence was ascending (e.g., 2, 7, 8). In the case of a sequence other than an ascending sequence, the investigator should refrain from reacting by pressing the button. As a result of the research, the following data were recorded: (a) simulator flight characteristics (degree of deviation of flight parameters from those prescribed in the instructions scored in points) and (b) completion level of the additional task (evaluation of sequences of digits). Individual differences in the level of flight performance were found. For example, in case of 16 people there was no response to the presentation of an additional visual task (sequence of digits displayed on the screen in the upper right corner); the respondents were only involved in piloting, although the level of performance deteriorated significantly compared to the flight without the visual distractor. On the other hand, in case of the remaining 80 people no such differences were found. In the case of an additional auditory task, there was no deterioration in the performance of an aerial task, although, as in the case of the visual distractor, there were also quite significant individual differences, which the authors explain with the influence of other psychological variables, such as emotions, temperament, etc. Although the authors did not confirm the basic thesis about the unambiguous negative impact of the additional task on the performance of the basic operator activity, they pointed out that observations are worthy of further research, because about 20 % of the tested people, despite the unambiguous instruction, did not perform the task of evaluating the sequence of digits, confirmed at least by pressing the reaction key, neither in the case of visual nor auditory presentation. Analysing the presented models exmplaining the behavior of operators in the situation of multitasking, it should be noted that they are complementary, it should be noted that they are complementary.

Thus, the first of the above hypotheses, explaining the course of performing two activities at the same time (*capacity-sharing* model), suggests that when more than one task is performed at the same time, the potential for data processing and responsiveness to more important tasks is divided. The authors of the aforementioned research claim that this concerns the behavior of the majority of their subjects. Although they also indicated that this more important task (piloting) also deteriorated. Therefore, the model with emphasis on *cross-talk*, which may also depend on other psychological variables, such as emotions, temperament, etc., is more appropriate to the research results obtained (Maciejczyk, Kuzak, Skibniewski, 1996).

B. Borowsky and T. Oron-Gilad (2016), referring to the increase in the number of automated road vehicles in the near future, conducted experimental research to answer the question: Will this new situation make vehicle operation easier or more difficult for drivers if automation fails and they have to return to "manual" control over their car? The research was carried out on a road vehicle simulator with specially prepared hazard scenarios, which were displayed on a curved screen. Situations which enabled the control of four modes of vehicle automation were tested: manual control without automation (M), adaptive cruise control (ACC), automatic steering (AS), and automatic driving (AD). Two types of additional tasks were taken into account: (1) road-situations related driving and (2) driving unrelated with road-situations. Eighteen drivers were tested, and they operated vehicles in the following order: (1) automated driving, (2) manual driving, (3) automated driving with additional tasks, and (4) manual driving with additional tasks. In each section typical hazardous events appeared (failure of automation and the need to take over manual control was alarmed by a sound and visually on the screen). The results showed that, although the involvement in an additional task not related to driving leads to more collisions, automation failure was not of much importance provided that the road was monitored by the drivers. However, significant individual differences in the operator performance of drivers during a secondary task not related to driving in automated driving mode were revealed. According to the authors, the explanation of these differences should be the subject of further research.

Multi-task interference is only one of the sources that disrupt the operator's visual attention. Other sources are shown in Fig. 6.

As Fig. 6 indicates, among the factors engaging visual attention and which are sometimes in interaction with each other (positive or negative) the following can be mentioned: (1) tasks requiring the processing of material with different modalities (visual-spatial-verbal), (2) switching between two processes,

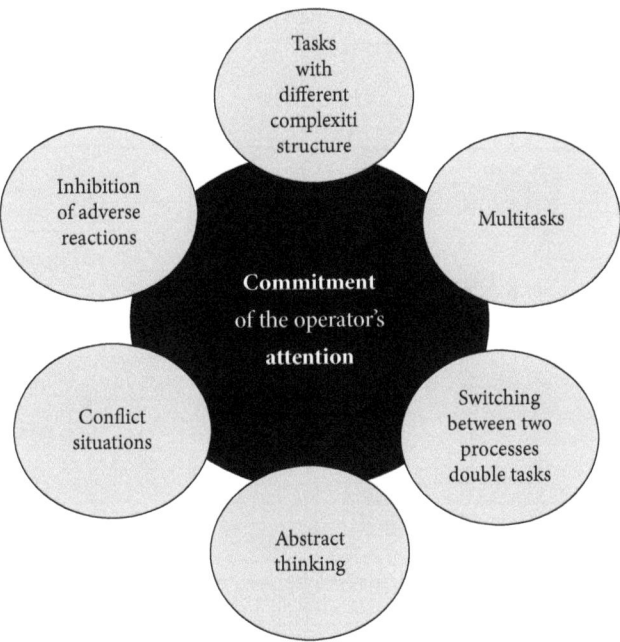

Fig. 6: Basic elements of the executive attention of the operator of technical devices (own calculations)

(3) abstract thinking, (4) conflict situations[33], (5) inhibition of adverse reactions, and (6) simultaneous performance of two or more tasks (see Fig. 5).

3.2.1 Role of visual attention in operator's spatial orientation

The ability to perceive direction and orientation in three-dimensional space is the basis for the survival and functioning of both animals and humans in the Earth's environment. When talking about directional orientation, although we do not always realize it, we always determine our position in three-dimensional space x, y, and z, for which the basic physical criterion of direction, coded in our genes, is

---

33 I am thinking here of the so-called Stroop Effect, experimentally confirmed in the situation of reading the names of flowers, which are written in a different font color than the natural color of the flower (in "The yellow flowers" the word "yellow" is written in a red font).

Psychological characteristics of the OMI system

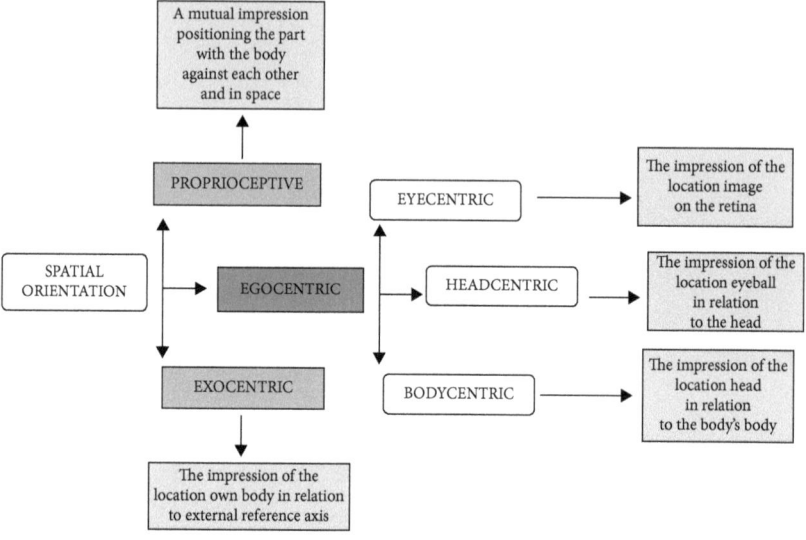

Fig. 7: Directional orientation systems (source: own elaborations)

the Earth's gravitation. The main axes x, y, and z, which determine the rotation of the human body, pass through the center of gravity of the body (in humans, near the navel). The "x" axis indicates a median-frontal plane, the "y" axis indicates a median-transverse plane, and the "z" axis indicates a median-arrow plane.

The separation of main axes and directional planes on Earth is the basis for orientation in the surrounding space for all living organisms. This applies both to the most primitive organisms, which react to changes in light with muscle spasms, as in the case of invertebrates (taxis), as well as higher order organisms, with various sensory organs and a developed nervous system. The structure of directional orientation in space is shown in Fig. 7.

In the case of human beings, three basic systems of directional orientation can be distinguished: proprioceptive, egocentric, and exocentric. These three systems are the basis of all human motor processes, while at the same time acting as a source of necessary information about the planned direction of action at the "input" and at the "output" – a control role, in the following three areas: *(1) Proprioceptive orientation* – is related to the evaluation of the relative position of individual body parts based on information from tendon, kinesthetic, and gravitational receptors (e.g., in darkness, under water, etc.). Commonly, this type of orientation system is called the "sixth sense" (e.g., when spinning on a carousel or using a lift); *(2) Eccentric orientation* – consists of three subsystems: (a) eyecentric, in which the field of vision is limited to an area of small angularly sharp vision (e.g., when the head is motionless); (b) headcentric, in which the field of vision is expanded in the range of the head's mobility; and (c) bodycentric, for which the characteristic field of vision is determined by the consequences of the location of the head in relation to the torso; *(3) Exocentric (allocentric) orientation* – this system is used when assessing the position of objects in relation to the external reference axes (e.g., the position of Kraków in relation to Warsaw). The latter system is of great importance, e.g., for pilots in the location of the airport or ship's captains at the time of entry into the port, etc.. Posturography is an example of proprioceptive orientation testing. A posturograph consists of a platform based on sensors, a monitor located at the eye level, and a computer interface, which receives digital indicators of the deviation from the projection of the center of gravity in the left –right and front –back axes. To evaluate the reactivity of the balance state and training of the vestibular organ, tests are conducted in three situations: with eyes open, with eyes closed, and biological feedback (observation of deviation from the vertical visualized on the monitor screen) (Kubiczkowa, 2000; Black, 2001).

An important intermediary variable in the O–M model is attention processes, without which a person cannot make an initial (input) selection of information useful at any moment for active action. Since the dawn of psychology, i.e., the end of the 19th century, attention was defined from the perspective of introspectionism by means of the *stream of consciousness* concept. Such a definition of attention still exists in colloquial language and can also be found in scientific studies, such as the classic monograph entitled *Experimental Psychology* by Woodworth and Schlosberg (1963), in which attention is treated as a factor that consciously selects one stimulus over another. It follows that in addition to being an active process of selecting stimuli, attention can be directed in two ways: *intentionally* – division to external stimuli or to the subject's internal activity or *detecting* when it is engaged (attracted) by external stimuli. In both cases, this was based on defining attention as a free and conscious process. Experimental research on attention, e.g., in the situation of recognizing ambiguous figures, evoking the illusion of the "entire object", extended the understanding of attention also to processes taking place outside person's conscious introspective control. Despite a number of rather chaotic fixations of vision, the image is cognitively interpreted according to the central "proximity principle", which in visual perception leads to the tendency of "the primacy of the whole over the parts". This is the reason behind many misunderstandings, as well as many deliberate actions, as in the case of masking or atavistic actions in animals, as in the case of mimicry.

Contemporary views are not only limited to the analysis of the content of consciousness (the stream of consciousness), but are also interested in the neuropsychological and psychological mechanisms of change of this consciousness at different levels of its organization. It is worth mentioning J. Konorski's physiological concept of perception (1969), in which, explaining the mechanism of instrumental conditioning, apart from consummatory reflexes (when a specific stimulus triggers a proper reaction, e.g., saliva secretion at the sight of food) and protective reflexes, he lists complex orientation reflexes, fulfilling an important cognitive role in a situation when the perceived information is new or important for the subject. The orientation reflex contains two components: (1) vegetative (nonspecific somatic and autonomous reactions, such as heart rate, respiratory rate, etc.) and (2) targeting, which is designed to adapt the receptor to receive the stimulus of a certain modality, and in the case of modality of vision includes, inter alia: eye accommodation, eye movements, and eye fixation pause. The psychological aspect of the mechanism controlling the targeting reflex in the field of human oculomotor cycles became the basis for the creation of contemporary theories of visual attention, among which a theory by the American psychologist M. Posner (1994) is worth mentioning. It can be described in general terms using the following theses, which form the basis of his theory of visual attention: (1) attention could be defined as a system that controls the flow of information and its importance; (2) attention management is the process that precedes the perception of a new stimulus; (3) noticing a new stimulus is related to the mechanisms of detaching attention from the previous stimulus. The author distinguishes three mechanisms of attention: (a) *orientation mechanism* – responsible for responding to new stimuli (so-called divided attention), (b) *detection mechanism* – allowing for intentional search for information (so-called focused attention), and (c) *maintenance mechanism* – allowing for continuous attention on one task (so-called sustained attention).

Thus, there are two basic mechanisms of visual attention management: (1) *orientation mechanism – overt attention* and (2) *detection or intentional mechanism – covert attention*. The basis for the division is the location of the stimulus controlling the attention (external vs. internal). It is assumed that both mechanisms are competitive with respect to each other (mutually weakening the effectiveness). This is evidenced, for example, by experimental studies on an operator performing two activities at the same time (Navon, Miller, 1987; Szczechura, Malawski, 1999).

These two types of visual attention mechanisms can be characterized as follows. *Orientation mechanism*: (1) works automatically, i.e., involuntarily in response to the appearance of external stimuli in the field of vision; (2) is responsible for the head's or the eye's movement in order to create optimal conditions for the fixation of vision on a cognitively appropriate fragment of the field of vision; (3) the speed of action of this mechanism in case of a sudden and strong stimulus is very high; and (4) is often accompanied by vegetative components (e.g., increased heart rate, breath, etc.). *Detection mechanism (intentional)*: (1) occurs in a situation of intentional searching for information in the field of vision; (2) filters one out of many possible stimuli; (3) stimuli that are not cognitively relevant at a given moment are omitted or at least received in a secondary way; and (4) the sought information may come from outside or be taken from own memory resources.

3.2.2 Oculomotor mechanism in visual attention processes

Returning to sharp vision of the so-called photopic vision, let us focus on an important fact of anatomical imperfection of the eyeball, which together with the system of further nervous connections ensures the highest visual sharpness located on a small area of the fovea centralis, covering the field of vision with a diameter not greater than 1.5 angular degrees. This means that only 1/10000 of the entire field of vision can be subjected to an accurate visual analysis at a time. This corresponds, e.g., to a range of 3–4 characters of standard printing seen from a distance of about 30 cm (Szczechura, Terelak, 1993), which of course differs from the real conditions of visual perception. The *scanning mechanism* created in the process of evolution, which relies on eye movements and fixations on particular parts of the field of view, during which information samples are taken and then transmitted to the central area of the retina, and in turn to the visual cortex of the brain, allows us to see sharply not only directly in an angularly small field of vision (*tunnel vision*), but also significantly expanding the possibility of sharp vision. We divide eye movements into small (involuntary, of a predominately physiological nature) and large (arbitrary) movements. The first group of so-called small movements shall be omitted, because they are not connected with intentional human behavior and have a physiological character, associated, among other things, with stabilization of the image in the retina (e.g., tremor)[34].

---

34 Analyzing eye movements from the functional side, it is possible to adopt specific types of eye movements: 1) atrial and proprioceptive reflexes (image stabilization on the retina), 2) rapid nystagmus phases (change of eyeball position), 3) tracking movements (keeping the image in the fovea centralis), 4) reflex and any saccades (change of fixation location), and 5) convergent eye movements (stereoscopic vision). Some difficulty in understanding the visual tracking mechanism results from the fact that about 1/3

Therefore, only eye movements that have a cognitive meaning in psychology, namely ankle movements and tracking, shall be discussed here.

Jumping eye movements (saccades) associated with the transferring vision from one object to another are connected with the change of the observed fragment of the field of vision and have a quantum and sequential character (i.e., consisting of alternating eye movements and eyesight fixation on a specific fragment of the field of vision). The measurement of such movements with eyetrackers is used in basic tests to describe the mechanisms of visual attention as well as in engineering psychology and ergonomics to design the perception field of various workstations, as well as in other areas of psychology, e.g., research on reading literacy, dyslexia, visual advertising, etc. (Jacob & Karn, 2002).

Pursuit eye movements – are connected with maintaining an object in the field of sharp vision when the object is moving in relation to the system of coordinates associated with the position of the head. According to Woodworth and Schlosberg (1963), they occur in natural conditions already after about 125 ms from commencing observation of a moving stimulus in the field of vision. The speed range of the visual stimulus, to which the tracking movement is adapted, ranges from a few angular minutes to about 40 degrees per second. Probably the mechanism responsible for tracking movement is a property of the retina and consists in a feedback loop reacting to the "slipping" of the visual stimulus from the retina area of sharp vision and alternating "transferring" the stimulus to this area.

Saccade movements studied from the perspective of cognitive psychology contributed to the objectivization of psychological theories of attention and memory (McCabe, et al., 2010). Knowledge of psychological mechanisms of involuntary (innate) and intentional (arbitrary) eye movements allowed visual attention theorists to move away from the classical paradigm of the "stream of consciousness", which has been in force since the beginning of psychology development, towards the knowledge of detailed mechanisms of cognitive attention. This is especially true for dynamic and static attention. Dynamic attention requires any tracking movements, i.e., following a moving object in the field of vision. As far as this type of attention is concerned, an interesting mechanism of involuntary transition to saccade movements was found when the eye is no longer being able to keep up with a stimulus as it starts moving with a velocity of

---

of people respond to the movement of their hands in complete darkness with eye movements. This means, according to Gregory (1971), that the central mechanism can strongly influence basic tracking reflexes.

about 30 angular degrees. In such a case, the eye makes a fast saccade movement from time to time, aimed at introducing and maintaining the object in the area of sharp vision for a certain period of time.

The oculomotor mechanism of attention is related to the egocentric directional orientation described from the position of *the eyecentric subsystem*, in which the field of vision is limited to a small angular area of sharp vision when the head is motionless. Although it is possible to extend the field of vision by moving the head or torso, at the registration level the direction of vision does not always coincide with the direction of attention, which became the basis of the model of visual attention by B. Fisher (1986).

The assumptions of Fisher's theory suggest, among other things, that when an interesting visual stimulus appears in the field of vision, but not important for cognitive purposes – then attention is detached and movement is made on the basis of involuntary attention. It is also possible to imagine any attention when looking for information important for cognitive processes (e.g., those provided by instruments) in the field of vision. A separate issue is the recognition and interpretation of information that are received by the receptors to a suitably specialized part of the cerebral cortex.

When the observer is warned where to expect a particular stimulus before being exposed to it, the attention transfer processes will follow a pattern that consists of operations to release the attention mechanism from its existing object of engagement (excitation, interruption, and engagement), to transfer the attention to a new area of the field of vision (covert attention orienting according to Posner, 1980) and involvement in a new area (involvement, blocking the return of attention). The main mechanism in this case is "orienting of attention", which does not have to lead to the perception of the object and the mechanism of "control over the orientation of attention". The latter mechanism can be controlled externally, e.g., by a strong visual stimulus or internally by an intentional act of search in the field of vision. The structural and quantitative characteristics of this cognitive cycle at the level of eyeball movements are presented in Fig. 8.

As shown in Fig. 8, the cognitive cycle at the oculomotor level includes the following components: (1) determination of the next location of eye fixation, (2) transfer of appropriate traffic information to the motor system in the brain, (3) eye movement to the new position in the field of vision, (4) sending a new visual impulse to the brain (CNS) for meaning analysis, and (5) decoding cognitive information contained in a new visual stimulus. Each of these elements has its own quantitative (ms) and content characteristics. For example, determining the next eye fixation (50 ms), impulse transmission (30 ms), eyeball movement (30 ms), information transmission to appropriate brain structures (60 ms), and

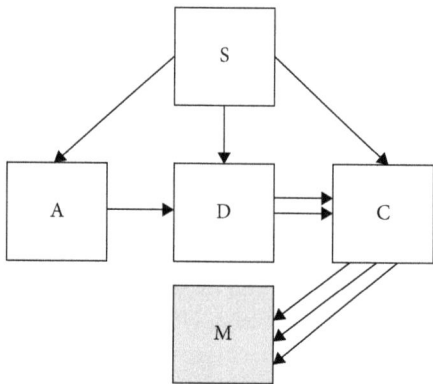

Fig. 8: Model of Posner's cognitive cycle at the eye-tracking level (own elaborations, based on Russo (1978)) (Legend: S – perception of the Stimulus, A – detachment Attention from the previous stimulus, D – Decision about eye movement, C – Counting eye movement parameters, M – eye Movement; Normal cycle – slow: S-A=D-C-M; Normal cycle – fast: S-D-C-M; Express cycle: S-C-M)

information decoding (60 ms). The entire cycle of the oculomotor process lasts 230 ms in typical conditions, and in stressful conditions it can be shortened to as little as 70 ms. From the point of view of content, a distinction is made between psychological processes preceding normal eye movement: the *Slow regular* cycle (S-A-D-C-M), which has a response time of about 200 ms, and the *Fast regular* cycle (S-D-C-M), which has a response time of about 140 ms. And *Express saccades* (S-C-M) with response times of up to 70 ms, which occur in stressful situations (Fischer & Weber, 1993). At the same time, normal eye movements (fast and slow) are arbitrary, while express eye movements are reflexes.

Therefore, saccades can be treated operationally as quantitative and qualitative indicators of cognitive processes at the level of visual attention. Precise determination of *eye fixation* times provides knowledge about individual differences related to the information content or cognitive meaning of a specific fragment of the field of vision against the background of the whole perceptual process. The order of fixation and its alternation determined by quantitative parameters creates the so-called trajectory of movements in the field of vision, because saccade movements are ballistic, i.e., pre-programmed, is no longer subject to any correction when begun. Thus, saccade eye movements are of great importance for studying the psychological mechanism of attention, as they involve the search in the field of vision for answers to three basic questions concerning the attention processes: (1) the location of the stimulus at a given moment, (2) the identification of the stimulus, and (3) the cognitive meaning of the stimulus. Various brain structures are involved in answering such questions, such as: superior colliculus (responsible for visual stimulus movement), visual field of the cerebral cortex (sensitive to the location and shape of the visual stimulus), and cortex of the frontal and occipital regions (responsible for targeting reflex and evaluation of the behavioral significance of the visual stimulus).

In operator's activity tests, two groups of problems related to the use of oculographic tests for the analysis of the oculomotor activity of the operator of technical devices are distinguished "at the input". The first group of research problems concerns the fixation of vision on sources of information about the state of the machine and the degree of implementation of the operator's task (instruments). This research is mainly of ergonomic nature, aimed at optimization of information instruments (construction of instrument panels). The second group of issues related to the examination of ophthalmic cycles concerns the identification of *scanning* patterns, which consists in "scanning" the instruments using the eyes, which is very important in engineering psychology to assess the progress of vocational training of technical device operators (e.g., when making progress during the training, the fixation times on individual instruments decrease).

The use of oculographic methods of objective registration and analysis of the operator's oculomotor cycles has been very promising in recent years for both application reasons (e.g., development of ergonomics) (Terelak, Szczechura, 1987) and theoretical reasons (e.g., development of attention and memory theory) (Nieznański, Obidziński, 2019).

3.2.2.1 Examples of basic research on the role of eye movements in visual observation

More specifically, the problems of simple visual cognitive behavior in the context of both voluntary (depending on intentional mechanisms) and reflex (depending only on the stimulus situation) elements of the visual attention system are the subject of my own research (Tarnowski, Terelak, 1996). According to a theory by M.I. Posner (1994), it was assumed that there was an influence of the decision-making situation on the functioning of the central mechanism of visual attention. It was expected that by introducing competitive stimuli, a change in the response time of saccade eye movements would be achieved only in the time overlap pattern. The empirical examination included two experiments with the use of OBER oculograph, which uses the reflection of infrared rays from the cornea. The response time of eye movement was measured in two experimental situations consisting in the presentation of central and competitive peripheral visual stimuli on a monitor screen: (1) the central stimulus was visible all the time (time overlap pattern, *overlap paradigm*) and (2) the central stimulus disappeared for 200 ms before the peripheral stimulus appeared (time gap pattern, *gap paradigm*). Each series consisted of a presentation of 30 stimuli. The decision-making process was controlled by means of peripheral stimuli (a number or a letter) appearing simultaneously next to a central stimulus ("cross"). Results of MANOVA variance analysis in the two-factor analysis model for dependent data with multiple measurement of the variable confirmed the assumed hypotheses that the decision-making situation did not affect the motor response time of the eye in the time gap pattern (*gap* =159 ms in a simple situation and 161 ms in the decision-making situation), while such an influence was observed in the time overlap pattern (*overlap* = 212 ms in a simple situation and 193 ms in a decision-making situation). Thus, the hypothesis on the impact of the decision-making situation on the central mechanism of visual orientation was confirmed.

Summing up the importance of the oculomotor mechanism of visual attention in contemporary theories of attention, it can be assumed on the example of Posner's concept that there are three mechanisms responsible for disturbing attention. The *orientation mechanism* is responsible for responding to new and sudden stimuli, and in particular for determining their location. Therefore, it is primarily related to the subsystem of binocular vision, so it has no intentional character. The *detection mechanism* allows for intentional, voluntary search for information. It is focused on conscious perception of visual stimuli and analysis of their characteristics, and as such it is controlled centrally. It is also very vulnerable to disturbance from other competing stimuli. The two mechanisms of attention discussed thus far are competitive in relation to each other, which means that people have a limited pool of attention resources (although not yet defined empirically). As a consequence, many additional stimuli (events) may be omitted if conscious attention is involved. Finally, the *retention mechanism* allows to keep attention continuously on one task or object. The detecting mechanism is therefore a function of the central attention mechanism, responsible for the overall control of the intentional action of a human being. For this reason, it is of great importance for the considered problem of distribution of operator's attention resources through a central mechanism (processor) capable to objectively record saccade eye movements and fixation pauses.

The positioning of Posner's Theory among other cognitive theories requires discussing the relationship between the concept of attention (especially the mechanisms controlling the direction of attention) and the concept of information processing, which are completely disjunctive. (Terelak, Tarnowski, 1999).

3.2.2.2 Application research on optimization of visual information sources

The analysis of the cognitive significance of the studied oculomotor cycles for optimization of visual information sources in the O-M system is based on the following assumptions, confirmed empirically in many studies of engineering psychology, namely: (1) the *frequency of vision fixation* on a given instrument is an indicator of its relative cognitive importance for a given operative task, (2) the *duration of fixation* (*Saccadic Reaction Time, SRT*) is considered an indicator of the relative difficulty of identifying and interpreting information from a given instrument, and (3) *the eyeball movements trajectory* in the field of vision is correlated with the functional links between instruments in a given operator task.

The empirical confirmations of the first assumption, concerning the frequency of vision fixation on a given instrument as a good indicator of its cognitive importance for a given operator task, are the results of own research (Szczechura, Terelak, 1981). The research subjects were jet plane pilots (n = 9 at the age of 27–35 years) flying on the KTS-4 simulator of supersonic aircraft in accordance with the ILS procedure (instrument landing system). The pilots were equipped with the Japanese NAC-V oculograph, which made it possible to record eye movements in the field of view horizontally – in the range of 60 degrees – and vertically – 43.5 degrees. Fixation points (light reflected from the cornea recorded using a miniature camera) were analyzed against the background of pilot-navigation devices and the total duration of fixation on basic pilot-navigation instruments: (A) attitude indicator (instrument informing about the spatial position of the aircraft in relation to the ground), (B) altimeter, (C) speedometer, (D) variometer (instrument informing about the speed of ascent and descent of the aircraft), (E) compass, and (F) fuel gauge. Six flight phases were recorded: (1) take-off, (2) heading turn, (3) climbing, (4) climbing turn, (5) descent turn, and (6) landing. The results recorded in 30,000 film frames were computer analyzed. Vision fixation frequency is determined by reading the absolute number of fixations per time unit. The research results are presented in Fig. 9.

As shown in Fig. 9, the most fixations concerned the attitude indicator, which also turned out to be the most visually controlled during the other five phases of the flight. This indicates that it is cognitively the most important source of pilot-navigation information in the entire flight. Similar results were obtained for the second indicator analyzed, namely the duration of fixation, which is considered as an indicator of relative difficulty in identifying and interpreting information

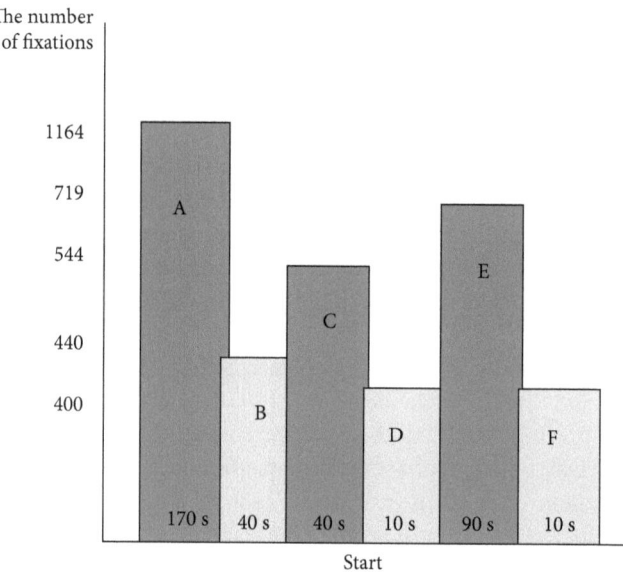

Fig. 9: Frequency of eye fixation and the proportions of the time of eye fixation (in sec.) on individual instruments pilot-navigating during the start (own elaborations)

from a specific instrument. This confirms that both analyzed oculographic indices of perception of the attitude indicator and the compass are highly correlated (r=0.90) and that in the case of the attitude indicator and the compass both the frequency of eye fixation and their total duration have relatively constant values. This means that they are cognitively significant in the entire flight, independently from the phase of flight performed. This does not apply to other analyzed instruments, the functional validity of which is closely related to the type of currently performed operator task (e.g., air combat) and the flight phase (e.g., take-off, landing, straight flight, etc.). The total fixation time on each instrument allows to assess its functional validity. For example, if a pilot fixates vision on the attitude indicator 20 times per minute, for about 0.65 seconds each time, 13 seconds per minute are spent on observing this instrument, which represents 22 % of the total visual perception time of all pilot and navigation devices. This information is of interest to ergonomists designing technical equipment workstations, which are controlled on the basis of instrument indications.

This type of research with the use of eye-tracking, assuming cognitive functional links between individual instruments, allows to determine the validity of individual instruments in the piloting process, the optimal location of which should take into account the so-called T rule (Green et al. 1996). It follows from this rule that the most important sources of information in the process of piloting an airplane, such as variometer, compass, speedometer, altimeter, and fuel gauge, should be located on the instrument panel in the central field of view at pilot's eye level. The legitimacy of this "T" rule is confirmed by oculographic studies, which provides important information to cockpit constructors (Jensen, ed., 1989). An example of such research can be own experiments carried out on behalf of the constructors of the new Polish helicopter "Sokół" based on another helicopter – "MI-2", designed for agricultural purposes (Terelak, Szczechura, 1987). The aim of the study was to assess the visibility from the first pilot's seat (left seat) of both individual pilot-navigation devices as well as visual information coming from outside the helicopter cabin. The latter are useful in the performance of various agricultural tasks for which the helicopter was designed, as well as in the performance of the only emergency maneuver available on the helicopter in the event of a technical failure, i.e., the "engineless autorotation". The pilot's eye movements were recorded using the Japanese NAC-V oculograph during real flight, including the following maneuvers: take-off, slalom, turn to target, turn to direction, agricultural turn, normal deceleration, turning deceleration, flying over an obstacle, rapid descent, "S" type maneuver, and autorotation. The visual fixation points (V-shaped marker) in the field of vision recorded on film tape were subjected to quantitative analysis using a special film reader coupled with a computer. Next, the points of fixation of the pilot's eyesight during individual maneuvers were graphically marked on a specially designed grid of rectangular coordinates, creating a standardized field of view from the eye level of the pilot sitting on the left seat. From the point of view of the pilot's visual perception, the following constructional defects of the prototype of what was to become the "Sokół" helicopter, hazardous to safety of flying in agricultural conditions, were found: (1) In the "take-off" maneuver, only one fixation concerned the indications of the "radar altimeter" and the rest were derived from extra-instrumental information, while the design of the right frame and the upper edge of the glass significantly reduced the visibility of the ground (in 36 fixations as many as 10 fell on those obscuring elements); 2) In the case of the slalom maneuver, it was also found that most of the necessary information comes from outside the helicopter cabin, which requires frequent shifting of sight in the whole field of view in the horizontal plane, and this is hampered by a poorly designed window frame; (3) The analysis of the trajectory of eye movements for the "agricultural turn" maneuver also shows that the orientation field for this maneuver comes from outside the helicopter (only two control fixations concern the "altimeter" and the "speedometer"), but some of them fall on the window frame, which is an obstacle. The field of vision beyond the cabin is wide, approximately 35 degrees to the right and left in a horizontal plane and the same amount above the central vertical plane of vision. However, during the field of view penetration outside the cabin, a significant number of fixation points were recorded on the windscreen frame, which clearly hinders visibility in the upper sectors of the windscreen of the natural horizon, especially during deep helicopter inclinations; (4) The analysis of the pilot's eye movements during the "S" maneuver showed, inter alia, that this maneuver is a typical example of visual orientation field penetration outside the helicopter's cockpit. The required visibility should be panoramic. However, the structure of the cabin windows presented for assessment has disadvantages related to, e.g., the helicopter's windscreen, which makes visual orientation difficult. This is evidenced by the sight fixation points registered under the lower, opaque window frame; (5) The analysis of the trajectory of eyeballs movements recorded during the "engineless autorotation" emergency maneuver showed so far presented structural defects of the cabin as well as new ones consisting insufficient amount of window space in the lower parts of the helicopter and overly large dimensions of the instrument panel. In this maneuver, the following instruments are observed alternately: speedometer, altimeter, compass, and altitude indicator with simultaneous switch focus to the natural horizon and the place of the imminent landing, located under the helicopter. The results of the research are illustrated by Fig. 10.

As shown in Fig. 10, structural limitations to the visibility of the ground were revealed in the case of engineless autorotation maneuver, which may pose a significant danger to the safety when landing the helicopter. On the basis of the analysis of the obtained results of oculographic research carried out for the needs of aviation ergonomics, it was recommended to aviation designers, among others, to miniaturize the instrument panel and design a panoramic windscreen for the helicopter, as well as to increase the area of side windows using very thin window frames and adding windows in the lower parts of the helicopter's floor.

Summing up the use of eye fixation indicators in engineering psychology research, a distinction should also be made between two types of eyeball fixation: control and monitoring fixation[35]. The basis for the distinction is the use of instrument-based information for a specific operator activity. Thus,

---

35 The well-known psychologist of visual perception L.L. Stark distinguishes two important types of seeing: "looking without seeing" and "looking and seeing" (Stark, Lis, 1981).

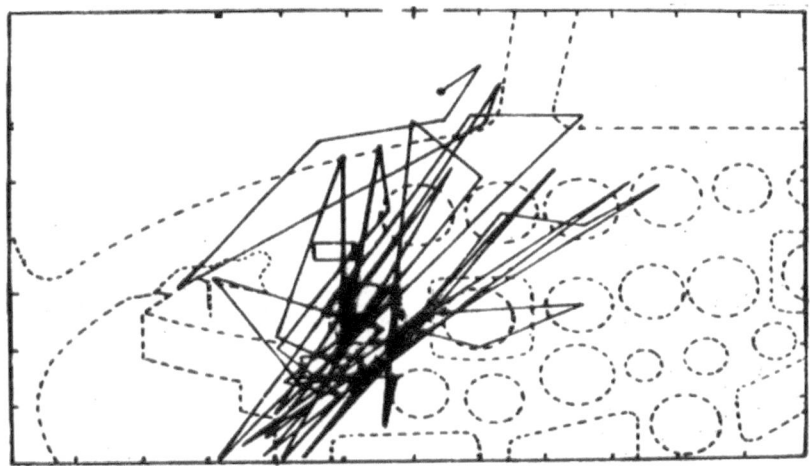

Fig. 10: The order and the part of the pilot's fixation of the pilot performing the "engineless autorotation" emergency maneuver on the MI-2 helicopter (own research)

control fixations are related to the direct use of information to introduce motor corrections in the machine operation, while *monitoring fixations* are only used to confirm, on an ongoing basis, that the planned and accepted state of machine parameters (operator action plan) is consistent with the factual state. In reality, it is difficult to separate the two types of fixation. The third aspect of research on eyeball movement concerns the possibility of outlining model trajectories (*scanners*) of eye movements in the field of vision, treated by engineering psychology as an indicator of functional connections of individual pilot-navigation instruments during flight in general as well as in individual partial (specialist) tasks of operators. These models are referred to in the literature on the subject as "information bundles". Their definition is not so much connected with working conditions, as with individual perceptual predispositions of an operator in terms of preferences of the sequence of observed instruments (Szczechura, Terelak, 1993). This is a problem of psychological differences between individual operators, used in the process of selection and optimization of operator training.

In addition to the application value resulting from the use of eye-tracking to assess the visual perception of operators, they are of great importance in basic research to verify the theory of visual attention. On the basis of previous research, psychologists agree that the oculographic tests can reflect the elusive mental process at the level of visual perception. The individual elements (fixation points) connected together give a certain image of eye movement in the field of vision, practically reflecting the real image on the retina. A good example of the thesis that the saccade eye movements are good correlations of cognitive processes is the classical oculographic study of chess players conducted by O.K. Tikhomirov (1976) using a photographic method. In these studies it was shown, among other things, that on the basis of recorded eye movements (trajectory) the so-called field of visual orientation can be determined. Although the author of this study characterizes the motor function of the eyes as a component of thinking, he does not identify them with perception processes, as he considers these movements to be exploratory activities with a high degree of exteriority. According to Tikhomirov, in the period preceding the final motor decision (e.g., the movement of a figure on a chessboard), the eye carries out a huge research work "combing through" all cognitively important elements in a specific situation in the field of visual orientation. This view was shared by a well-known psychophysiologist of visual perception, I.M. Sechhenov (1986), as early as the 1940s, who claimed that "the human eye can be compared to a hand that learns about an object through 'touching'". This thesis has been confirmed empirically by contemporary researchers that the motility of the eye is of a cognitive nature and not mechanical. The L.L. Stark's concept is interesting in this respect. It explains the mechanism of motor visual patterns recognition, known in the literature as the concept of a "ring of features" (Stark, Lis, 1981). This concept assumes, among other things, that during the recognition of visual patterns, the observer performs a repetitive sequence (*scan path*) of alternating saccade movements of the eye and fixation pauses, creating the so-called ring of features. This ring of features consists of alternating sensory elements of the eye, which are the observable features of a given image in the field of vision (during the next fixation pause) and the motor elements of the eye (another saccade movement), which are equivalent to the structural organization of the field of vision. For example, the oculographic research on facial recognition based on vision fixation points shows that *face recognition* is largely based on the concentration of fixation in a triangle of relations – "eyes – nose – mouth". Other parts of the face are less cognitively important. Nowadays, this oculographic pattern of faces appearing on the Internet in millions of copies per day in the form of *selfies*, is used by new, controversial *facial recognition* technologies, in the so-called personalization of emotional and motivational states, by, e.g., "Amazon" (*Amazon Recognition*) or "Microsoft" as commercial products, addressed to a variety of commercial users.

Another example of the use of oculographic studies can also be the inclusion of temporary oculomotor parameters of a pilot to assess the process of flight training, both in terms of the difficulty of the aviation task (information load level indicator) and in the various stages of training. This is the subject of separate research with the use of a version of the Japanese NAC-V oculograph, conducted on the simulator KTS-6 for the supersonic aircraft Mig-23 (Szczechura, Terelak, Kobos, Pińkowski, 1998). In these studies, it was found that the analysis of the oculomotor dynamics of changes in the reading of instruments during flight is useful for objective assessment of the increase in workload. With an additional task in relation to piloting, consisting in the use of a visual distractor (reading a series of randomly generated digits, appearing in the field of vision) and an auditory distractor (repeating and sorting the digits given to the handset), it was found, among other things, that eye fixation times on pilot-navigation instruments and fixation pauses, associated with the need to change the strategy of reading the instruments, were prolonged.

The data presented thus far, concerning the possibility of recording eye movements, are of an application nature. This data also contributed to the development of research on the role of visual attention in the evaluation of sensory and motor skills of operators.

3.3 Psychomotor mechanisms of operator's action and measurement thereof

The nervous system does not only regulate the processes of reception and processing of information inflowing to our senses, but also coordinates executive decisions using systems responsible for energy and motor aspects of human life functions. Such systems include the skeletal-articular-muscle system, which creates a biomechanical system that controls human motor skills at the level of both psychomotoric abilities and motor skills (Osiński, 2000). The subject of our interest are not so much the motor skills, but above of all the operator's movement skills included in the cognitive process related to directional orientation and coordination in three-dimensional space.

As can be seen from the characteristics of directional orientation systems presented earlier, they are very important in the psychomotor behavior of an operator. The psychological mechanisms of these systems can be considered as sequences of orientation processes consisting in tracking visual (auditory) information with parallel so-called motor compensation, which aims at minimizing the error between the set state of a machine (e.g., the position of instrument indications assumed in the device manual or maintaining the machine in an appropriate spatial position) and the current state (current position of instrument indications and its spatial position) (Moray, 1999). This inclusion of motor activities in the cognitive process is a sufficient justification to treat these activities as an integral part of the cognitive process and call them *psychomotor activities*. The basic range of psychomotor activities of the operator includes simple sensomotor reactions (e.g., response times) and motor habits, as well as complex processes of visual-motor coordination.

3.3.1 Sensomotor reactions and motor coordination

Modern technical devices with a high degree of automation make it much easier to perform motor tasks, sometimes limiting them to switching various types of levers, knobs, or buttons on or off. However, such movements require selective attention (selection of the lever) and deliberate motor reaction (Raczek, 1993). This is related to the preceding perceptual and decision-making processes, which are responsible for a properly executed motor response. The speed of reaction is connected with individual differences, which are quite significant and are the typologically conditioned properties of the nervous system, as well as with the strength and modality of the stimulus (e.g., response time to light is 180 ms, to sound, 140 ms, to touch – 140 ms) as well as the state of the sensory organ, attention, skill, etc. (Woodworth, Schlosberg, 1963, Vol. 1, p. 73). The response time, which consists of the time of sensory activity, the conduction of nerve impulses, brain activity, and the time needed for the relevant muscles to start acting, is measured in total by means of special devices called chronometers (chronoscopes).

Concepts explaining the cerebral mechanism of reflecting reality are quite diverse. From a functional perspective, the brain consists of three different but functionally integrated systems. The first is related to the reception, processing, and transmission of information (nervous mechanism of sensory processes); the second regulates cortex tension and the state of awareness (physiological mechanism of activation); while the third one programs, regulates, and controls complex forms of psyche (nervous mechanism of mental processes). These blocks do not function in isolation, but perform an integrative brain function, i.e., they cooperate in organizing and coordinating all forms of behavior, due to which an animal or human receives information about the external world and its body's internal states, as well as reacts to perceived changes (Konorski 1969).

Three phases can be distinguished in the structure of the sensomotor reaction: A + B and C1 + C2. The first two A + B, known as the *latency period*, lasts from the moment of the signal's initiation (e.g., visual or auditory stimulus generation) to the beginning of a motor response. The third phase consists of the dynamic part – C1 (characterized by high speed and fluidity) – and the stabilization part – C2 target motion (includes correction micro-movements around the target). The third phase is therefore the correct measure of the duration of movement. In the case of motor habits, the stabilizing phase practically disappears, while in the initial stages of learning a motor activity, the dynamic phase is not very clearly marked. In practice, the measurement of simple and alternative response times, recorded using a variety of chronometers, reflects all the phases mentioned (Uszakowa, 1969).

The average time of psychomotor response to a stimulus with different modalities can be formulated as follows: **CR = f (B, O)** (where: CR – response time; f – function; B – stimulus of a specific modality; O – persons) and vary depending on the modality of the stimulus. When examining the reaction time, it should be remembered that the choice of modality depends on the experimenter, and that the studied person may influence the final result (e.g., fatigue, individual differences in temperament, level of motivation, etc.).

From the neuropsychological perspective, in the temporal aspect a motor reaction can be divided into three basic components: (a) *Nerve impulse conduction time from a receptor to the appropriate nervous system center (CNS)*. The time of nerve conduction depends on many variables, such as: physical strength of the stimulus, receptor sensitivity, quality of the nerve fibers that conduct the impulse, and functional state of the synapses that switch the nerve impulse from neuron to neuron. The last three variables depend on the overall physiological reactivity of the body and the state of activation of the CNS; (b) *The time of motor organization in the specialized motility structures of the CNS*. This time varies according to the structure of the stimulus. The more complex the stimulus, the longer the time. Experimental studies on the time of complex (alternative) responses have shown that the alternative response time increases with the number of choices. This dependence is known from the literature as the *Hick's Law*, which can be expressed using the following formula: **CR = K log2 (n + 1)**, *where: n – number of alternatives, K – constans squared*. The formula indicates that when signals with the same probability appear, the response time is proportional to the number of alternatives. A large number of experiments confirmed Hick's hypothesis that the rate of receiving information is more or less constant at about 5 bits per second. This law is used to study the scope of attention, which is indicated by "wrong reactions" and, in practice, in so-called masking; (c) *The time of nerve impulse conduction along nerve paths descending* to specific muscle group. In experimental studies on the time of simple psychomotor reaction to an auditory stimulus, the respondents reacted in four series (50 bopeds) with right and left hand and left and right leg. It was found that there were no differences in simple reaction times between left (144 ms) and right hand (147 ms) and between left leg (179 ms) and right leg (174 ms), while there were differences between the reaction times of hands and legs. Legs, being further away from the CNS, have a longer time of nerve impulse conduction along descending pathways to specific lower limb effectors (Woodworth and Schlosberg, 1963, Vol. 1, p. 73).

3.3.2 Psychological mechanisms of motor reaction speed

When discussing alternative response times, based on the electrophysiological mechanism of the excitable cell's functional potential, it was assumed that the next stimulus always appears after the end of the previous motor response. And what happens if there are further stimuli coming at short intervals before the previous motor reaction ends? It was experimentally found that in such situations the response time to the next signal is longer than the previous one. This phenomenon has been described as a *psychological refractory period* (following the pattern of electrophysiological refraction[36]), which is that the next stimulus, provided within a too short time after the first stimulus, cannot start a new motor response, until the motor response to the first stimulus is completed. However, the second stimulus appears at the time of the motor response to the first stimulus. Thus, the speed of response to the second stimulus depends on the value of the time interval that separates it from the first stimulus (refraction is 0.5 s). Moreover, it was empirically demonstrated that psychological refraction also occurs in the case of eyeballs fixation (Szczechura, Terelak, 1993).

The psychological perspective is based on the assumption that the stimulus, besides its formal structure (strength, modality), is also characterized by a content structure, as it is the carrier of specific information. Experimental studies on the speed of subjectively received information indicate the need to focus not only to the suitability of the stimulus and reaction, the degree to which the motor reaction can be trained and the degree of differentiation of stimuli, but also to the comparison of the dependence of the speed of psychomotor reaction on the amount of information the stimulus is the carrier of. As far as the first three issues are concerned, it was stated, among others, that the dependence of the time of simple reactions on the skill exists, but it is insignificant and achievable not after a few attempts, but after several hundred, spread over time even of over

---

36 By the term "electrophysiological refraction" I mean the property of a cell that is periodically insensitive to stimuli that stimulate it after passing the action potential, which may manifest itself in two forms: the cell is not able to respond to any stimulus (absolute index of refraction); the cell is able to respond to a stimulus greater than physiologically intensity (relative refraction).

several days. According to experimental data, for example, in the case of the time of a simple reaction, the increase in skill after one day is only 10 %, and in the case of an alternative reaction time 30–40 % (Woodworth, Schlosberg, 1963, Vol. 1, p. 67). It was also found that the reaction time depends on the age of the examined person, which assumes the function of a developmental curve (similar to the training curves), characterized by the fact that up to 25 years of age the reaction time decreases and then slightly increases.

In adulthood, it remains more or less at the same level, up to the age of 60 and then slowly begins to increase, but generally at a slower rate than we observe in terms of visual-motor coordination. The acceleration of a simple motor reaction under the influence of motivation (reward and punishment) was also observed. Concerning the dependence of the psychomotor reaction rate on the size of the information, which is carried by the stimulus of a specific modality, it was experimentally indicated that there is a connection between the time of the alternative reaction and the amount of information. As indicated by the *Shannon's* analysis, the psychological equivalent of the term "amount of information" should be considered in two ways: (a) as the *signal probability*[37] (degree of signal uncertainty, degree of unexpectedness) and (b) as the *number of possible signals* (degree of complexity of choice). Both of these psychological parameters determine the speed of psychomotor response differently, as evidenced, among other things, by other information processing operations. The strategy of using the *probabilistic structure of a stimuli sequence* correlates with the distribution of response times to both the stimuli which occur seldom and frequently. In this case, response time is an objective criterion for the subjective probability of a stimulus occurrence, and the strategy itself consists in optimizing the process of receiving information by increasing the speed of signal reception, at the expense of receiving frequent stimuli. This clearly shows that even at such a simple level of psychomotor processes as reaction times man is an active subject, effectively solving motor tasks, and not a static recipient of stimuli. The regulator of people's activity in this respect is the "significance" of psychological (content) stimuli, which causes an increase in the speed of psychomotor response.

---

37 The term "signal" is understood in psychology as an abstract model of different measurable quantity, changing in a certain time, described mathematically (formally) and in terms of content (information carrier). Sometimes it is identified with a "stimulus" that triggers the reaction of one of the sensory organs.

90 Psychological characteristics of the OMI system

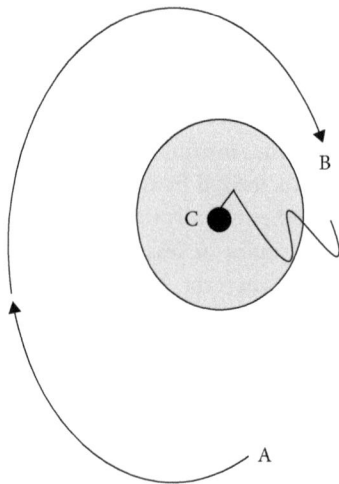

Fig. 11: Mechanism of sensory correction of targeted movements (own study based on: Nazarow (1969) (A – the beginning of the movement, B – the starting point of the stabilizing correction, C – the final point of the movement)

In terms of behavioral mechanisms of movement speed, human beings particularly distinguish themselves among other primates due to the huge number of degrees of freedom available to the human motor apparatus, especially the hands. The mathematical model of the hand's motility includes only the initial moment of its movement and is based on a single muscle. Due to the enormous complexity of the human musculoskeletal system, further stages of the hand's motor process have so far become an unsolvable task for robotics. The control of such a complex human motor apparatus, apart from the cerebral kinematic chain, would be ineffective if it was not for the mechanism of sensory correction of targeted movements, presented graphically in Fig. 11.

The second important regulatory mechanism of human motility at the behavioral level is the *mental image of movement*, which far surpasses its behavioral form, anticipating not only the motor act itself, but rather the effect of comparing the actual and preset values of parameters in the context of the performed motor task (e.g., following an escaping moving target by moving the airplane's yoke) (Chapanis, 1969). Both these mechanisms play an important role in shaping sensomotor coordination.

Theoretical considerations on the mechanisms of sensomotor response time should be concluded with the mention of the use of quantitative reaction time parameters in psychological diagnostics. It is believed that the most common use of the measurement of the psychomotor response time is: (1) a description of the state in which *certain nervous system centers* are located (e.g., studies on the influence of verbal stimuli on the inhibition of response times, thus establishing the existence of the previously mentioned psychological refractory period); (2) a description of the general state of fatigue (e.g., it was found that sensomotor reaction time becomes prolonged under the influence of sleep deprivation and some pharmacological agents); (3) determination of typological differences, conditioned by the type of the nervous system (temperament); (4) determination of the rate of perception of information by people (e.g., evaluation of reactivity of senses and ergonomic value of signaling scales and indicators); and (5) determination of the degree of differentiation of modality of stimuli. The above also suggests that although the sensomotor response time is not a universal measure of "optimality" of conditions, its measurement can be successfully used as an additional indicator of various aspects of a simple psychomotor process. The process of sensory-motor coordination is much more complex. It should be considered as a sequence of orientation processes, consisting in tracking the visual stimulus with the so-called motor compensation, aimed at minimizing the discrepancy between the position of the visual stimulus (target) (e.g., a stochastically moving cursor on a monitor screen) and the actual position, by means of corrective movements. According to the classic of American engineering psychology – A. Chapanis (1969), *human tracking behavior* is a kind of "Skinner box" in a zoopsychology laboratory. The basic research paradigm for the tracking process is the *visual to steering wheel ratio* (e.g., rotary indicator – rotary steering wheel, linear indicator – rotary steering wheel, rotary indicator – linear steering wheel, linear indicator – linear steering wheel, complex indicator – rotary steering wheel, complex indicator – control stick). These relations are conditioned by certain types of motion acts, the so-called operator acts, which include: (1) *Position movements* – consisting in changing the position of a part of the body (e.g., changing the position of the hand when the appropriate lever is moved, etc.) and including such typical movements as: single movements, strictly limited to the motor operating field: (2) *Continuous movements* – forced by changes in machine parameters and related to the operation of control devices (e.g., stick, pedals). The direction, speed, and range of such movements depends on the degree of discrepancy between the current dynamic state of the machine and the prescribed operating parameters; (3) *Serial movements* – a separate, independent series of movements carried out by the operator in accordance with the machine's operating procedure (e.g., engine start procedure in a car, parking procedure, etc.).

3.3.3 Psychological mechanisms of coordinated sensomotor movements

Sensomotor coordination depends on human directional orientation systems, and in the case of operative action – on the egocentric orientation system. It is used when it is necessary to evaluate the position of an object located in space (e.g., a switch, rod, button, etc.) in relation to the body (corpus-centric system) or in relation to the head or eyes (headcentric and eyecentric systems). Two subsystems of egocentric orientation can be distinguished, depending on whether the reference axes are fixed in relation to the observer or are mobile. If the axes are fixed, we are talking about a *pure egocentric orientation* (e.g., when following the command to position yourself parallel to the wall). If the references are part of another external system (e.g., compass direction), we are talking about a *semiegocentric system*. Its essential feature is the relative stability of orientation, regardless of the position of the observer. The operator's psychomotor space, determining various motor characteristics, is characterized depending on the type of directional system (Kroemer et al., 1988). For example, in the operating space of the cockpit of a passenger aircraft pilot there are three zones, taking into account the biomechanical properties of humans: (1) a zone with instruments situated in the central field of vision, (2) a binocular vision zone, and (3) a zone requiring movement of the head and torso.

The model by I.P. Howard (1982) is useful for a detailed analysis of the operator's visual-motor coordination in a specific multidimensional space. The model in a simplified form is presented in Fig. 12.

The model of visual-motor coordination analysis presented in Fig. 12 is based on a system of egocentric orientation. This model consists of three important elements: sensory, skeletal, and muscular, and is limited to the horizontal plane and only to the motility of the left hand. However, it should not be forgotten that the number of components is much higher in the real operation and the complexity of their cooperation is much higher. In line with Howard's model, the operator task requiring visual-motor coordination can be described as consisting of the following components: (a) the determination of the target of motion as the modal point from which hand-to-target deviation is estimated; (b) determining the center of the eye rotation from which the direction of the pupil is measured; (c) determining the center of the head, which is the basis for the position of the hand in space; (d) a description of the deviation of the head (C) and hand (D) from the median plane of the body; (e) an indication of the angle of deviation of the target in relation to the body (E); and (f) an indication of the angle of deviation of the axis of vision from the median plane of the head

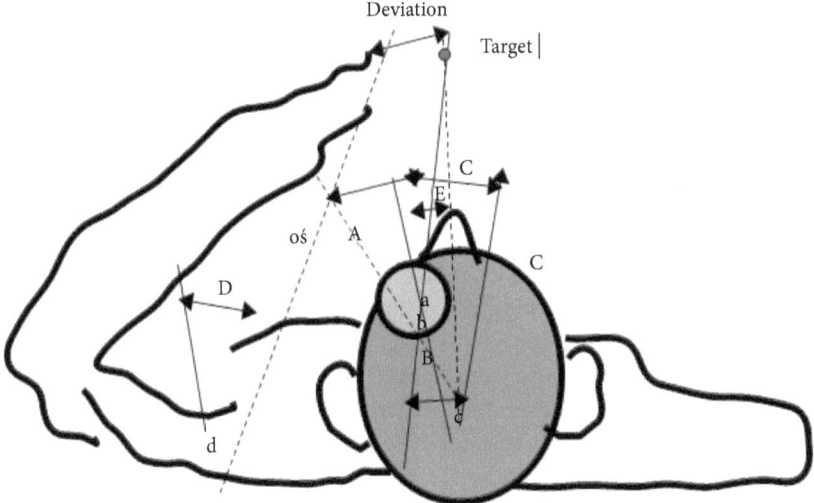

Fig. 12: The procedure of analysis of the left-hand coordination structure from the eyecentric perspective (own calculations based on the Howard model (1982))

(A), which is the angle of deviation of the target from the axis of the eye. If we assume that the mentioned components are coaxial, we can assume that $E = C - (A + B)$, which means that the angle of deviation of the target from the body (E) is the value of deviation of the head (C), decreased by the sum of parameters of the angle of deviation of the target from the axis of vision (A) and the angle of deviation of the axis of vision from the body's median plane (B). As a result of this the operator, at the brain level, is able to assess the position of the target in relation to the body's centerline (E), calculating information about A, B, and C. Such a calculation is the basis for making a decision and sending it to the motor part of the cerebral cortex in order to program an appropriate corrective motion in the task of visual-motor coordination. An appropriate sequence of such calculations is necessary in the performance of those operative tasks in which the position of the target changes dynamically, forcing subsequent mental motor programs, consisting in the reduction of discrepancies between the target position and the current one. Therefore, the level of visual-motor coordination depends on three factors simultaneously: (1) the level of accuracy of the calculation of the corrective motion, (2) the speed and accuracy of the motor act, and (3) the constant readiness for motion in conditions imposed by the task. The latter factor is especially important for understanding the dynamic aspects of visual-motor coordination, as it makes us aware that the motor act is not a transition from "immobility" to "movement" but is included in the human cognitive process. This is evidenced, among other things, by "preparatory motor activation" and sometimes incorrect motor reactions of the "a-ha" effect, which in psychology are referred to as the level of alertness (readiness to act) (Terelak, 1994).

Let us now focus on the speed and accuracy of single movements. They depend on several factors, such as the direction of movement, its range, and interaction with other motor activities. These relationships are defined by concepts such as the duration of feedback processes and the duration of motor corrective acts. This is not the case with motion sequences, in which direct feedback is replaced by appropriate brain motor programs. Usually, operators do not deal with a single motor act but with motion sequences. Bearing this in mind, it is worth recalling some data concerning psychomotor activity of a human being in general.

The quickest movements that can be classified as class A are those that are stopped by contact with an object. Class B includes the movements that are characterized by a low degree of precision (e.g., the movement of the hammer stopping after bouncing off of a nail). The slowest movements fall into class C, they require the highest degree of precision (e.g., grabbing and moving objects). The duration of movement also depends on whether it takes place under visual supervision or not. In the latter case, the time of the execution of the motor act increases significantly. The *accuracy of movement* can take the following trigonometric relationship: $y = a - b \cos 2x + c \sin 2x$ (where: y – accuracy; x – direction of movement; a, b, c – constant). *Movement execution time* is a common function of both range and required motion accuracy. This dependence is known as *the Fitts's law*, which states that movement time (CR) with twice as large range will not change if the size of the target also doubles. This can be presented using the following mathematical formula: $CR = a + b \log_2(2A/W)$ (where: A – motion amplitude; W – target width; a, b – constants). Thus, the velocity characteristics of some motor reactions vary greatly due to the size of the target, which determines the motor act. However, regardless of the quantitative and qualitative characteristics of motor acts, it should be stated that both at the level of the brain (motor programs) as well as the behavioral one (motor acts) they are subject to training, creating motor habits.

The most typical of any operator movement is the *complex visual to control stick* relationship, especially if the control stick has dynamic characteristics, including control of position, velocity, and acceleration along two coordinate axes (x, y) with a two-dimensional compensation tracking system. In this way, it is possible to create a model for testing visual-motor coordination with different degrees of task difficulty. Among other things, the literature indicates that the tracking efficiency with motor compensation is best when the control method along both axes (x, y) was the same. As divergences in this area increase, there are errors indicating an increase in the difficulty of a psychomotor task. Furthermore, it was found that in one-dimensional tracking (visual signals moving sinusoidally), the visual-motor coordination was the best when controlling speed and slightly worse when controlling acceleration and the worst when controlling the position of the object (Łomow, Platonow, 1984).

More complex tasks requiring psychomotor coordination are assessed by means of specially designed simulators of operator working conditions similar in many respects to the real ones. In addition to assessment as a criterion for selection of candidates for various types of operator activities, simulators are very helpful for learning new work activities and their stabilization in the form of habits.

4. Learning new operator activities

Vocational training of operators covers both the acquisition of knowledge concerning the operation of a technical device as well as the training of practical skills, in particular the consolidation of occupational habits. Since detailed discussion of individual operator groups exceeds the scope of this study and a rich literature is available on the subject (Schultz & Schultz, 2008), this study shall focus on the practical aspects of the training of new operator activities and working habits on the example of the profession of a pilot, who controls the most technically complex machines. This imposes the highest level of operational requirements for a modern candidate for aviation work, both at the recruitment level and the further stages of selection due to the necessity of aircraft training of new generations and new tasks (Tsang, Vidulich, eds., 2003).

Tony Smallwood (2000), passenger aircraft captain and pilot training consultant at the London Heathrow Airport and lecturer at *Oxford Air Training School*, in his monograph *The Airline Training Pilot*, draws attention to the importance of vocational training of aircraft operators. In his opinion, the vocational training of pilots should include both the theoretical and the practical part. The theoretical part concerns both the permanent expanding of knowledge about state-of-the-art computer-controlled electronic devices and modern methods of passenger ship crew management. One of the basic tasks of the vocational training of an aircraft operator is shaping: (a) sensomotor habits (automated activities) useful for efficient control of a technologically complex machine; (b) intellectual capabilities both in terms of operational thinking and anticipation of events occurring very dynamically, as well as in terms of quick and accurate decisions in conditions of threat to health and life; (c) adaptive mechanisms of functioning in dynamically changing situations, such as time limit, threat to life, being responsible for the safety of others etc. and (d) raising the psychophysical fitness needed to cope with extreme working conditions (e.g., high vs. low temperature, hypoxia, load factor, weightlessness, vibrations, ionizing radiation, etc.). The three main types of vocational training techniques, which are most often used in practice, are: (1) simulation of simple or complex operator tasks, (2) modeling of single or combined working environment factors, and (3) real-world training (Patrick, 2003).

4.1 Simulation of simple and complex operator tasks

The basic ability of people to adapt to the requirements of the working environment results from intelligence, the essence of which is learning new behaviors. Learning operator's activities shall be understood as a process leading to the modification of previous professional behaviors, which in the phase of well-established habits shall lead to an optimal O–M relationship, being a criterion for the assessment of occupational suitability in a specific line of work. According to T. Tomaszewski (1968), professional suitability can be illustrated by the following formula: **Pz = f (K, M)** (where: Pz – professional suitability; f – function; K – qualifications; M – motivation to act). The term "professional qualifications" is understood by the author as a combination of abilities (general and special) and vocational training, which is the result of accumulating professional experience and learning not only the rules of action but also specific practical skills. The general characteristics of the learning process is covered in rich literature on the subject (Buskist, Davis, 2007).

The issue of learning habits and operator skills is the subject of theoretical psychology (psychology of learning) and applied psychology (psychology of work). It should be remembered that the operator's habit is a process or a sequence of psychomotor processes learned up to their automatic execution. The concept of habit is understood by cognitive psychologists, not so much as the automated activity itself, but rather the acquired ability (disposition) to perform this activity. Disposition, being more complex and less automated than a habit, is a skill, characterized by readiness to take a given type of operator action, with the ability to adapt it to changing working conditions (e.g., flying at a low altitude, flying through clouds, air combat, etc.). Sensomotor habits and skills refer to activities in which the musculoskeletal system plays an important role under the constant control of receptors, but a key role is played by integrative brain activity, generating mental programs of future operator actions (Konorski, 1969). Observing the process of developing sensomotor habits reveals many interesting relationships, e.g., between the number of repeated psychomotor activities and learning outcomes. Different typical courses of gaining proficiency, known in the literature as *learning curves*, are characterized by such features as: (a) *positive acceleration* – characterized by an initially slow and then rapid increase of the skill level until it gradually reaches the *plateau*; (b) *negative acceleration* – i.e., an initially fast and then slow increase of the skill level until it reaches its optimum; and (c) *mixed system* – reflects a difficult beginning and slow increase of the skill level followed by a rapid increase and decrease of the skill and its stabilization at a not too high level of learning (S-shaped learning curve). These classic learning curves do not exclude others, such as learning in stages, where the increase in proficiency is marked cyclically at an increasingly higher *plateau* level (Widerszal-Bazyl, ed., 1998). In practical terms, it is possible to plot individual learning curves empirically, because apart from general regularities of learning, its effectiveness and style is also reflected by other anatomical determinants (e.g., efficiency of the motor apparatus) and psychological determinants (e.g., type of nervous system, efficiency and modality of receptors, motivation, task difficulty, etc.). As far as the latter determinant is concerned, namely the difficulty of the operator's task, it should be remembered that the speed and accuracy of individual movements also depend on several factors, such as the direction of movement, its range, and interaction with other activities (the so-called dual tasks involving the senses of different modalities, e.g., sight and hearing) (Terelak, Tarnowski, 1999). These relationships determine the duration of feedback processes and the duration of corrective changes. However, the issue of performing motor sequences presents itself quite differently. In this case, direct feedback is replaced by appropriate global mental motor programs. This applies to the work of many operators of complex machines, such as ships, aircraft, spacecraft, whose work is related to psychomotor sequences with a high degree of difficulty of visual-motor coordination and a high level of proficiency. These psychomotor sequences can also be divided into several classes of movements depending on the speed of their execution. Generally speaking, three classes of such movements can be distinguished: A, B, and C. Class A includes the quickest movements that are stopped by direct contact with an object (e.g., catching a ball). If movement is stopped by antagonist muscles (class B, C), the time of movement execution increases. Class B movements are characterized by a low degree of precision (e.g., the movement of the hammer stopping after bouncing off of a nail). The slowest movements fall into class C, and they require the highest degree of precision (e.g., moving objects). The duration of movement largely depends on whether it is performed under visual control. If there is no visual control or it is limited, the movement time shall be noticeably extended. This is because all movements (simple and complex) require mental motor programs. They are based on visual and kinetic information received by the brain regarding distance from the target which must be converted into a series of detailed commands sent to appropriate muscles. Then, when the movement is initiated, it is kept under the control of such a program and cannot be changed until some time elapses, so the brain can process feedback information about the effects of the previous movement. The concept of mental psychomotor programs becomes more important when it concerns a series of task movements or proprioception (spatial orientation), as there is a high correlation between the visual system and balance. Thus, learning sensomotor activities makes a lot of sense, as it perpetuates cerebral visual-motor programs through their mental internalization. Only then do they have the regulatory power in the form of psychomotor habits. Their effectiveness depends on the amount of practice spread over time, taking into account the difficulty of the task (simple vs. complex motor acts).

Sensomotor habits are practiced using various types of simulators of operator's work and actual training flights with an instructor. This issue shall be analyzed on the example of learning psychomotor activities by an aircraft operator, which is part of vocational training. Generally speaking, every vocational training covers two aspects: the acquisition of specialist knowledge and the training of specific professional habits and skills. The principles of acquiring knowledge are well described by theorists of learning, on the one hand they include the development of work habits in laboratory conditions, and on the other hand, in conditions similar to the actual requirements of the working environment. The former are the result of using different types of simulators.

4.1.1 Simulators for learning simple structure sensomotor tasks

Simulators for training the ability to perform simple structure operator tasks, called "coordinometers" or "trainers", cover a whole range of static equipment, such as video games (Hull, Draghici, Sargent, 2012) up to analogs of various road, air, sea, and space vehicles (Anderson et al., 2005).

An example of a simulator to used study a simple structure of sensomotor tasks of pilots from the 1970s of the 20th century can be the SMA-3 coordinometer produced by Bryans. It was a quasi-electronic device consisting of two components: an oscilloscope screen on which a visual stimulus controlled by a hand (a stick) and legs (pedals) is presented. When the program is switched on, a spot of light with a diameter of about 2 mm appears on the screen of the oscilloscope, which moves along the screen for 90 seconds according to a randomly generated trajectory. The operator's task is to maintain the spot of light in the central position, in a square of 4 cm x 4 cm, by means of horizontal (using pedals) and vertical (using a stick) corrective movements (so-called compensatory movements). It is a model task in an imposed work pace, evaluated on the basis of such quantitative parameters as: (a) the number of corrective movements of arms and legs, the number of times the spot of light leaves the central area of the screen, and the time during which the spot is maintained in the central area of the screen. The SMA-3 simulator is presented in Fig. 13.

On the basis of own research with the use of the simulator presented in Fig. 13, it was found that the learning curve of visual-motor coordination activities for

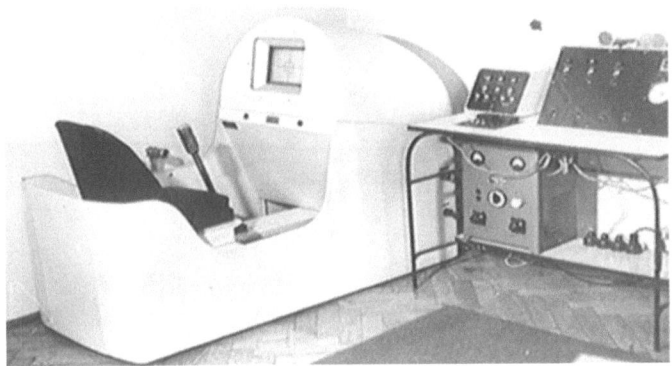

Fig. 13: The 1970s Bryans SMA-3 static simulator for examining visual-motor coordination (own archive)

pilots stabilizes (plateau) after 8–10 training sessions (Terelak, 1995). Moreover, significant individual differences in the learning pace were found, conditioned by many psychological variables, among which the type of the nervous system, stress, as well as perceptual and motor skills seem to be at the forefront. For example, let us recall two experiments on stressful and perceptual conditions for effective learning of tasks requiring visual-motor coordination. Own research included 20 pilots aged 25–40 years tested on the SMA-3 simulator in normal and stressful situations (500 mA electric shock administered to the temple by means of a battery as a penalty in case of improper action). It was found that under moderate stress, better learning outcomes (difference in first and eighth sample measurement) are achieved. However, they come at a higher psychological cost (at the heart rate measurement level).

In other studies using the same SMA-3 simulator, differences were found in the learning process of sensomotor activities of candidates for pilots (N=25) conditioned by cognitive styles correlated with types of visual perception patterns, whose oculomotor structure is depicted in Fig. 14.

As shown in the Fig. 14, two oculomotor structures of operators, conditioned by the so-called cognitive styles: conservative vs. anticipatory, were distinguished empirically. As we know, the level of performance of the task related to visual-motor coordination depends on the mechanism of anticipating the position of the visual stimulus on the monitor screen. This skill allows for a relatively quick reduction of the discrepancy between the actual position of the light stimulus and the ordered one. During the process of learning of visual-motor coordination the

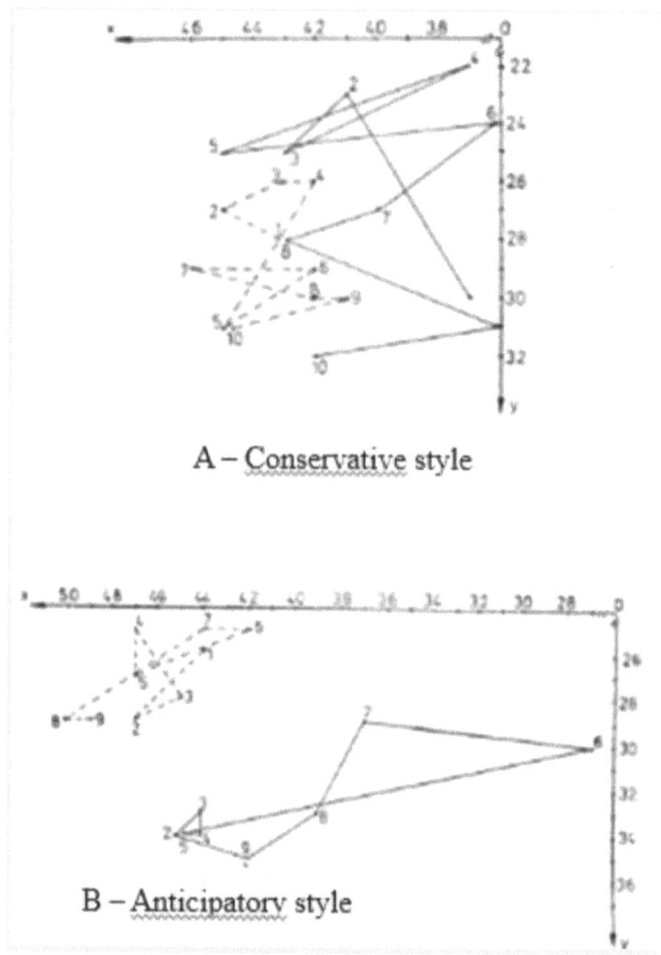

Fig. 14: The structure of the trajectory of eye movements reflecting individual differences in cognitive styles: conservative vs. anticipatory (own research)

eye movements were recorded using the NAC-V eye tracker, the trajectory of eye motility and eye fixation points were reproduced on the oscilloscope screen. It has been assumed that oculomotor indicators are behavioral equivalents of the anticipation mechanism. A correlation was found between the rate of learning of visual-motor coordination and the pattern of visual perception. Better anticipation (at the level of eye movements trajectory) allows for a relatively quick reduction of the discrepancy between the anticipated position of the light stimulus (the spot of light on the oscilloscope screen, the equivalent of which is the registered point of fixation of vision) and the ordered position (4 cm x 4 cm square in the center) (Szczechura, Terelak, Świątek, 1988). The diagnostic accuracy of these tests was confirmed with the flight instructors' assessment, who evaluated the progress in real flight training using a specially designed sheet. The study was carried out as follows: on the basis of the instructors' assessments, three groups of pilot school students were identified: good, average, and mediocre (25 persons each), which were tested using the SMA-3 simulator. The test model consisted of eight repetitions in accordance with the standard program within two days (four tests each day). Differences were found in teaching a task requiring visual-motor coordination in the three examined groups in favor of the group designated by the instructors as "good". In addition, a coefficient of multiple correlation was calculated, from which it appears that the sum of tests 1 + 2 + 7 + 8 ($r = 0.55$) is the most prognostic. This means that individual differences in teaching tasks requiring sensomotor coordination using simple coordinometers can be a basis for forecasting the effectiveness of practical pilot training. This also applies to all types of simulators, which are more often used to train complex operator tasks (Migdał, Terelak, 1986).

4.1.2 Devices simulating operator's physical working conditions

Theorists of aviation training agree that there was an important breakthrough in aviation training due to the possibility of using various types of simulators in the process of learning operator activities, reproducing more or less faithfully the actual working conditions and tasks to be performed. Referring to the history of flight simulator development, it should be mentioned that the first such simulators resembling real flight cabins are associated with the name of the American aircraft designer Link (Green, Self, Ellifritt, 1995). This type of flight simulator, commonly known as the "Link Trainer", is shown in Fig. 15. For many years, it was used for psychological tests at the Military Institute of Aviation Medicine in Warsaw.

The development of flight simulators is in line with advances in aviation technology, aviation psychology, and aviation ergonomics. The idea of using simulators in aviation training is just as relevant today as it was at the beginning of their creation, particularly since due to the development of microcomputers it is easier to create a virtual version of dynamically changing flight conditions. However, there is no consensus on their classification and didactic usefulness. Flight simulators are generally divided into three groups: (a) stations

104   Learning new operator activities

Fig. 15: The 1960s Dynamic Link simulator used to test the pilot's effectiveness (own archive, courtesy of the Archive of the Military Institute of Aviation Medicine)

reproducing single or combined future flight conditions; (b) simulators designed for teaching and developing operating habits; and (c) combined methods, skillfully combining operator tasks with extreme flight conditions. The first group includes training stations and means designed for physical fitness preparation and specialized preparation (e.g., special gymnastic and exercise equipment, etc.) as well as simulators imitating the physical environment of an airplane or spacecraft (e.g., a centrifuge, decompression chamber, a low vs. high temperature chamber, etc.).

The second group includes simulators for training professional habits (e.g., navigation, communication, landing or water landing, air combat, etc.). The third group of simulators includes different combinations within both groups mentioned above.

In addition to specialized and functional simulators, comprehensive simulators (meeting the criteria of specialization and functionality) and universal simulators are used to train pilots or astronauts. The latter can only apply to airplane-laboratories, which together with the instructor have the task of consolidating the previously acquired operator habits on various types of simulators. This type of training is designed to consolidate skills in different flight phases and tasks. All flight simulator training is conducted until the moment when operator habits are performed automatically, in all flight situations, including stressful situations (failures, night flights, instrument flights, air combat, etc.). The last stage of, e.g., aviation training takes place on the ground in a dynamic model of a given type of aircraft or in a training aircraft, which is the most accurate analogue of working conditions, both in terms of the tasks performed and the combined action of the so-called physical flight factors.

4.2 Cognitive models of operator's activity

From a psychological perspective, the essence of modeling operator's activities in working conditions similar to real-life ones is based primarily on the creation of a mental image of the future action, which is a derivative of a specific objective. This objective is a regulator of the future performance. In cognitive psychology, the goal of a specific action or logically related sets of actions is considered as its ideal, imagined, anticipated result, i.e., as something that does not yet exist, but should be achieved in the course of action due to the goal set beforehand. Due to the visual character of the so defined goal, it is called "image-goal" in the literature (Terelak, 1988). Being an ideal reflection of the future result, "image-goal" functions as a premise defining its origin. No operator activity related to the performance of a specific task (objective) can be planned or specified without such a premise. For example, in aviation practice, the "image-goal" of a specific aviation task is defined in the "Flight Operations Regulations" (*Pol. RWL – Regulamin Wykonywania Lotów*), which formulates the objective of the task under precisely defined conditions. The Flight Operations Regulations shall cover all past aviation experience, including knowledge of the means of action and the state of the subject of the action. This is very useful for shaping the mechanism of anticipation (attitude, extrapolation) which itself is useful in taking individual actions in a specific order and conditions. The results of experimental aviation psychology research confirm that the "image-goal" based on the mechanism of anticipation determines the criteria for selecting information about the state of the object of action, as well as its synthesis, which is the basis of the developed operator's working habit (Łomow, Płatonow, 1984). Thus, what information from the general data flow will be selected first and foremost by the operator and in what order it will be used depends on the "image-goal" of the future action, as this image also determines the ways in which the incoming information is decoded, evaluated, hypotheses are formulated, and decisions are made regarding the operator action. Achieving an "image-goal" is a continuous process, taking place in a defined time sequence and in a defined space, realized through detailed tasks, each of which is carried out by separate activities. For example, in the case of an air force pilot, the "image-goal" associated with air combat is not to perform difficult maneuvers associated with take-off and landing, but primarily to detect another aircraft or rocket in the airspace, maintain radio communication with other aircrafts, observe instruments, intercept enemy aircrafts, initiate an air fight (practice or indeed), report on the performance (or nonperformance) of the task, and return to the home base airfield. If, after T. Tomaszewski (1963), according to his "Theory of Activities", let us assume that the element of activity connected with the performance of one simple current task is an activity in a psychological sense, then the whole task formulated in the "image-goal" can be described as a system of changing activities in a specific time and space. The "image-goal" formulated at the beginning of the action should be stored in the operator's memory until its completion, acting as the regulator of the whole system of actions. In the case of established working habits, the "image-goal" should be easily reproduced from memory when needed. However, the system of changing operator's working activities may vary in structure and, according to Tomaszewski's views, may create a normal (easy) or stressful (difficult) situation for the operator. In the former case, the action does not require a significant change in the structure of the action, as it may be characterized by a predictable consequence of the action, where each action prepares the next, which is its logical extension (algorithmic level of habits). In the latter case, when an operator has to carry out several tasks relating to the same objective simultaneously or has to achieve several objectives which arise unexpectedly (e.g., air combat strategy depends to a large extent on the opponent), the structure of the activity must be modified according to the difficulty of the task or the achievement of new objectives (heuristic level of the activity). An example of the need to change the structure of activities was the Moon landing performed by Armstrong according to the "manual" rather than automatic procedure, because the lack of data on the topographical conditions of this astronomical body caused a faulty computer evaluation of the landing site, which would be on the side of the volcanic crater with a high slope, could lead the lander tipping over and tragically end off the entire lunar mission. However, regardless of the complexity of the structure of operator activity, an "image-goal" should be stored in the memory throughout the durations of the task's execution. It is a difficult task and requires appropriate theoretical training and operational preparation, which in the psychology of work is defined as a learning process aimed at shaping mental models: conceptual and operational (psychomotor). These models result from the "image-goal"; they are a mental representation of the projected tasks and the means needed to achieve them. The achievement of this goal is served by various forms of vocational training as well as training with the use of various types of simulators of operator's work. Both aspects of vocational training should include both knowledge and the ability to adapt to the requirements of the working environment. Only the objectives of vocational training will be considered here, which includes, among other things, the shaping of: (1) sensomotor operations up to the level of automated operations; (2) intellectual properties (conceptual model of operation including, i.a. operational thinking, event anticipation, etc.); and (3) psychological properties (ability to make decisions, cope with threat stress, control anxiety, take risk, etc.). The three base types of training techniques, which are most often used in practice, are: (A) creation of a conceptual model of work, (B) simulation of work activities and working conditions close to the real-life ones, (C) adaptation of a person to the requirements of real working conditions. In the case of pilot's vocational training, these techniques include, but are not limited to, flight conceptualization, flight simulation, and performing actual flights with an instructor.

4.2.1 Conceptual models of operator's activity

In order for the operators to carry out their task, they should not only have a global idea of the future result ("image-goal"), but also obtain current information on the changes taking place in the controlled technical object. This involves a mental process based on the reception and processing of information into appropriate decisions made by the operator. As is known, many operators do not have the possibility to directly observe the changing states of the controlled technical object (e.g., electric locomotive, aircraft, nuclear power plant, etc.), therefore the necessary information is received on an ongoing basis from various indicator systems in coded form. As it was explained earlier in Chapter 1, the reception of information is a process that takes place at two levels: sensory and mental. The first one involves the perception of physical phenomena occurring as carriers of information (e.g., position of the instrument needle, light brightness or color, sound intensity, etc.). The second – mental – concerns decoding the received signals and formulating on this basis the mental plan of the process of controlling the technical object, taking into account the changing conditions in which this process takes place. Such a mental plan of purposeful action came to be known as a conceptual model of an operator's task in the psychology of work. Such a model usually covers not only individual activities in an algorithmic (habitual) way, but also their relations with the general objective of action at each stage of its implementation, including situations of failure of a technical object and/or disturbances of the subject of action. In the shaping of the conceptual model of work, an important role is played not only by the direct perception of situations, objects, incoming stimuli, but also by a system of meanings encoded in memory (e.g., language, pictograms, etc.). The latter property is successfully used in vocational training that prepares operators to work with a specific technical object in the future (e.g., vocational schools, vocational training and retraining, ending with certificates) or in training prior to specific operator tasks. An example of such training in military aviation are, for example, various techniques used in so-called ground-based flight preparation, aimed at creating a conceptual model of a specific aviation task. In aviation psychology, a mental flight model is understood as an image of one's position in three-dimensional airspace, which was created under the influence of past aviation experience and external real conditions. This image sometimes does not coincide with reality, leading to the so-called aviation illusions, i.e., incorrect interpretation of the situation and taking improper action, which is the cause of many aviation accidents (Maciejczyk, Biernacki, 2005). In such situations, operating memory, shaped in the process of acquiring aviation knowledge and personal experience, proves to be useful. One of the elements of enriching aviation knowledge and personal experience is ground-based training preparing pilots to perform new aviation tasks, including detailed discussions about all flight phases, graphical representations of its course, and spatial presentations of individual states of position in a three-dimensional space. This last element of ground training in aviation takes place on airplane models and special gymnastics exercises, among which particular attention should be paid to exercises called in aviation jargon – "flying on foot" and aviation gymnastic instruments (e.g., carousel, Rhönrad – gymnastics wheel, etc.). Let us stop for a moment and discuss gymnastic techniques as a way of shaping a conceptual flight model. As far as the first "flying on foot" technique is concerned, which consists in the pilot making various turns, leaning forwards and backwards to the sides, I must admit that years ago when I was a beginner aviation psychologist, it did not impress me at all. I saw a group of grown men running in the hall, with arms stretched out and performing various "movements". Initially, I found it amusing. Only after years of experience in aviation psychology did I appreciate these exercises and their "wisdom". Let us recall that the pilot and the plane become one in a three-dimensional space. The technical possibilities of airplane movement in three-dimensional space are far greater than the experience of human species, because this space is not our natural environment. Without an innate mental model of the dynamics of an airplane in three-dimensional space, people have to introduce into the brain unitary experiences related to the training of their own sense of balance and proprioceptive system, which are responsible for assessing the position of the body in space. The lack of such individual experience increases the risk of misjudgment in real flight dynamics, as the actual information on the spatial position of the aircraft, in the absence of information in the memory, may be misinterpreted. My many years of experience in the Aviation Accident Investigation Committee (*Pol. Komisja Badania Wypadków Lotniczych*) confirms the above thesis about the link between the lack of individual experience (especially in case of young pilots) with the correct assessment of the spatial position of one's own body and aviation accidents caused by the loss of the spatial position of an aircraft (Terelak, Truszczyński, 2000).

The second more dynamic group of gymnastic exercises are exercises performed on special Physiological Training Devices, including, e.g., looping, Rhönrad, gyroscope, carousel, etc. Their usefulness in the field of ground-based aviation training was confirmed, among others, by research conducted by J. Terelak and Z. Kobos (1996). The authors examined the visual-motor coordination (Düfoure apparatus) of 80 students in aviation school before and after 30 hours of training on Physiological Training Devices (looping, Rhönrad, and gyroscope) and found a statistically significant improvement in psychomotor coordination. This is in line with views of J.P. Guilford (1988), who claims that psychomotor coordination is subject to learning and its effectiveness depends, among other things, on the interaction of three factors: limb coordination, spatial orientation, and visual imagination.

Although the conceptual model formulated in the course of ground-based aviation training is very useful in a real flight, its abstractness and conceptual level makes it insufficient to create a detailed image of the future aviation task. It is therefore necessary to plan not only the general direction of operation, but also the individual steps. An operational model of the operator task, including simulation of work activities and working conditions close to real-world conditions, is useful for this respect.

4.2.2 Operational models of operator's activity

As it was mentioned earlier, an action is carried out as a result of successively performed work activities, each of which is responsible for performing a part of the operator's task and achieving a partial result (goal). Each separate work activity is subordinate to the general objective of the operator's task and is related to those activities which have already been carried out as well as to those which are still to be carried out. Proper relations between activities subordinate to one goal are possible only if the operator has an action plan organizing these activities in a specific time and space. Such a plan, shaped in the mind even before any specific action is taken, is called in the psychology of work an operational model of action, which takes into account both the acquired work activities and working conditions. Training on a simulator is very useful in this respect.

4.3 Training of operator's action in variable working conditions

There are specialized devices, which help to get used to the physical working conditions, simulating: high and low temperature (thermal chambers), high and low atmospheric pressure (decompression chambers), and linear and angular accelerations (high-G centrifuges, ejection seat simulator). The suitability of some of them for aviation training will be discussed below.

4.3.1 Training in thermal chambers

The impact of temperature on operator's efficiency can be summarized as follows: (a) at a relative humidity of 50 %, full ability to work is achieved at a temperature of thermal comfort of about +20 degrees Celsius; (b) slight difficulties at work (annoyance, irritation, difficulty in concentrating, mental performance reduction, etc.) are noted in the temperature range of +21–26 degrees Celsius; (c) significant discomfort at work (increase in the number of mistakes, decrease in dexterity, increase in the number of accidents, etc.), raising the temperature to 27–30 degrees Celsius is associated with an increase in temperature to +32–35 degrees Celsius; (d) physiological dysfunctions (water-salt metabolism dysfunctions, significant strain on the cardiovascular system, heavy fatigue, risk of emaciation) and the accompanying decrease in heavy labor efficiency appear in the range of +32–40 degrees Celsius; and (e) the highest bearable limit temperature, accompanied by significant physiological dysfunctions, is over +40 degrees Celsius.

The analysis of the impact of thermal stress on the human organism is usually preceded by the assessment of the so-called thermal comfort, determined on the basis of a subjective assessment of the microclimate conditions of the surrounding environment. Such conditions should not cause any intensified natural defense reactions, such as sweating or shivers, in the organism. Dysfunctions of subjective thermal comfort are accompanied by changes in physiological functions of the system, associated with the functioning of the thermoregulatory mechanism. Research conducted by J. Łaszczyńska (2002), to name one, shows that the training carried out in thermal chamber reduces the period of adaptation to working conditions at elevated ambient temperature and increases the operator's awareness of the need to replenish fluids, as the loss of heat at elevated ambient temperature disturbs the liquid-electrolyte balance (the so-called dehydration) and acid-base (sodium loss) balance. In turn, heat exhaustion caused by dehydration is characterized by a decreased ability to work, general weakness, and exhaustion of the body. The heat exhaustion caused by sodium loss is characterized by headaches, apathy, nausea and vomiting, general weakness, lack of appetite, calf muscle cramps, and even circulatory collapse (Radakovič et al., 2007). Lack of training in thermal chambers resulted in the loss of ability to work and disruption of many functions of the bodies of operators who did construction work in sugar factories, power plants in Libya and Iraq years ago. Medical data and occupational psychology show, among other things, that in the case of working steelworkers, miners, etc., who have adapted to the described climatic conditions, we are dealing with biological selection, not adaptation to these conditions (Blatteis, 2007). For the purpose of such selection, decompression chambers can be successfully used as thermal simulators.

A more detailed discussion of the stressful nature of high temperature (hyperthermia) and especially low temperature (hypothermia) shall be omitted, as this problem was the subject of chapter 2.2.1.1.

4.3.2 Training in a decompression chamber

As is known, proper functioning of a human being is possible within the limits of atmospheric pressure of 800–660 mmHg (1060–880 hPa), typical of pressure fluctuations on the Earth, because they do not change the partial oxygen pressure in the body significantly. However, the problem appears when ascending above sea level with the subsequent reduction in atmospheric pressure and when pressure decreases below sea level. In both cases, human functioning reduces the ability to work and sometimes endangers life (in so-called critical areas). In order to increase the ability to diagnose one's own body under conditions of changing atmospheric pressure, the operators who have to work under conditions of reduced atmospheric pressure (e.g., pilots, astronauts) or increased pressure (e.g., divers) undergo adaptive training. Special simulators called decompression chambers are used for this purpose (Kowalski, 2002). Different terms are used in the literature of the subject to describe the state of insufficient oxygen supply to the organism. The most commonly used term is "hypoxia"; this word of Greek origin means reduced oxygen content (Reinhart, 2008). The definition of high-altitude hypoxia highlights the phenomenon of physiological and functional dysfunctions, therefore all trainings in the decompression chambers are based on these two aspects of determining the individual level of tolerance of hypoxia and possible increase in the level of adaptation of the body. The results of the studies conducted so far indicate high plasticity of human adaptation mechanisms to conditions of hypoxia in terms of cognitive processes and physiological reactions. Data found the literature on the subject indicate that hypoxia has a significant impact on conscious activities, and to a small extent impairs automatic activities (Molińska, 2015).

Cerebral hypoxia also occurs in case of deep-sea divers in the event of failure of the oxygen tank or people diving without oxygen tanks (e.g., pearl divers). A critical zone begins at about 50 m of immersion, which, in addition to the loss of ability to work with acute states of physiological insufficiency, is accompanied by states of decreased self-awareness (so-called nitrogen narcosis or "raptures of the deep"), which precede the loss of consciousness and death (Vaernes, 2007). The objective of training in decompression chambers is learning how to recognize the body's states preceding the loss of consciousness.

4.3.3 Training in a high-G centrifuge and a catapult

Accelerations, which are a consequence of a sudden change in speed and direction of flight, result in dysfunctions of the body's function caused by the force of inertia (Reinhart, 2008). Using the standard acceleration due to gravity of 9.81 m/s^2, the acceleration of a jet plane can be defined as a multiple of gravitational acceleration. The influence of accelerations on the human body is tested in specially constructed high-G simulators, called *high-G centrifuge* or *centrifuge*, which are used to determine the acceleration tolerance limit and to assess the operational efficiency and to train to increase the tolerance of accelerations, especially impact accelerations, especially during using catapult and air combat (Walichnowski, 2008).

What is more, according to *Newton's second law of motion*, it is possible to calculate the value of the force applied to the body if its weight and acceleration are known. The most important factor for the efficiency of pilot's functioning is training in terms of centripetal acceleration (+Gz), during which the weight of pilot's body increases multiple times, e.g., at 5+Gz the weight of pilot weighing 80 kg, equals 400 kg (with retained mass), which has physiological (e.g., cerebral hypoxia) and psychological consequences (deterioration of psychomotor coordination and sometimes making it impossible to perform targeted movements), which affects even the well-trained operator's performance (Wojtkowiak, 2015). The most dangerous disorder caused by acceleration is +*Gz-Induced Loss of Consciousness*, which is a hemodynamic disorder occurring at the head level, a symptom of which is cerebral hypoxia, manifested in three phases: initial *grayout*, then *blackout*, which is preceded by a *tunnel vision* (Albery, 2004).

Prevention of the *G-LOC phenomenon*, besides using anti-g suits, involves increasing acceleration tolerance in three ways:

(1) Comprehensive training (physical exercise, isometric, anaerobic, and speed training).
(2) Training in the high-G centrifuge according to two programs: (a) *Gradual Onset Rate (GOR)* (increasing accelerations linearly at a specified rate until the moment of complete loss of peripheral field of vision within the limits of 50 degrees); (b) *Rapid Onset Rate (ROR)* (increasing accelerations at a specified rate in cyclically repetitive intervals separated by short breaks).
(3) Special training, carried out in real training and combat flights, allowing to accumulate individual experience of the impact of accelerations on operator's efficiency.
(4) Individual training of breathing technique under conditions of so-called respiratory hypertension, aimed at reversing the natural passive phase, spontaneous exhalation of air from the lungs in the direction of active exhalation, which is important in conditions of accelerating action, exerting enormous force on the pilot's mediastinum, making breathing in a natural rhythm difficult. Currently, both US NAVY and NATO pilots of high-maneuver airplanes are required to obtain a special certificate in the field of breathing-related anti-G straining maneuver, called the "HOOK maneuver"[31], which in the case of accelerations consists in the ability to perform energetic exhalation (every 3–5 seconds), simultaneously uttering the word "HOOK", with the glottis partially closed, pronouncing "HOO" within a maximum of 1 second, and ending by vigorously shouting "K", thus closing the entire HOOK word (Winter, 2001; Wojtkowiak, 2004).

Another simulator used for learning the conditions of rapidly increasing acceleration in the situation of emergency lowering the altitude of an aircraft is an impact catapult (Wojtkowiak, Szajner, 2000).

A separate problem related to accelerations are aviation illusions called *somatogravic illusions*. These illusions are a result of stimulation of otolith organs by linear accelerations and are based on false subjective sensations of a spatial disorientation character, that a person exposed to linear acceleration induced by the Earth's gravitation has the impression of the direction of the force resulting from the acceleration as if it was caused by the Earth's gravitation. The most dangerous illusions for pilots are the somatogravic illusions, as there are no methods of training resistance to them, apart from raising the level of awareness that during the flight one should not trust their senses but the navigation instruments, which for this purpose are duplicated and supplied from various sources by aircraft constructors (Terelak, Truszczyński, 2000).

Summing up the usefulness of the discussed simulators, one should agree with the statement that simulators of partial working conditions are a good analogy of real work, but only in the scope of developing tolerance of physical factors of the working environment, as they do not include the structure of work activities focused around the task objective, carried out by individual operators. For this reason, operator activity trainings are used in complex and universal simulators enabling operator work activity exercises.

4.3.4 Training of operator actions in comprehensive simulators as analogues of work experience

Comprehensive simulators and universal comprehensive simulators used for training of operator actions are stationary and can come in static (computer visualization) and dynamic (e.g., imitation of movement in space) versions.

The universal simulator (although the name itself is misleading, because it suggests that it can be used to train operators working with different types of machines) is a specially adapted training "laboratory" in which, under the instructor's control, it is possible to acquire and practice operating habits related to functioning in different conditions (e.g., at night, during the day with visibility, in fog, etc.) and tasks (e.g., parking, overtaking, emergency braking, etc.). Therefore, comprehensive simulators need to be an analogue of individual types of machines and vehicles, airplanes, ships, etc., so they need be equipped with appropriate steering and navigation control devices, visualization of the visual perception field, and computer simulation programs for different variants of tasks. For these reasons, training on both types of simulators is an expensive undertaking (Tichon, 2007). As research conducted by, among others, Ch. T. Scialfa et al. (2013) shows, this cost is related to the visualization of the dynamic conditions of simulated car driving; the comparison of the results of a study of beginner drivers on visual perception in static conditions (paper-pencils tests) and dynamic conditions (a passenger car simulator) resulted in a not very high correlation of 0.40.

For example, a dynamic car simulator is made up of a driver's cab, an instructor/operator's station, a visualization system, a motion system, a surround sound system, an IT system, an analysis station, and provides the possibility of an accurate evaluation of the task at hand. Because of that, this type of simulator has a wide spectrum of application, such as: testing the predisposition to becoming a professional driver, training of candidates for drivers, and improvement training in driving in special conditions (e.g., in mountainous terrain, in winter conditions, in metropolitan traffic, in left-hand traffic, with an unusual load, etc.). From a psychological standpoint, training on this type of simulator allows to master vehicle control skills (e.g., visual-motor coordination), prediction of the development of the situation on the road (e.g., anticipation and spatial orientation processes), and reaction to stress (e.g., during car breakdown or difficult road situations, etc.).

A number of experimental studies on dynamic and multi-task simulators confirm the thesis that training sessions of aircraft operators positively correlate with the level of performance of aviation tasks in real flight. In own research carried out on the simulator KTS-4 Yak 40 (it is an analogue of a small passenger jet airliner – Yak-40), the correlations of operation of pilots in the situation of simulator training and in real flight were analyzed (Szczechura, Terelak, Świątek, 1988). The KTS-4 Yak-40 flight simulator consists of the following elements: (a) a cabin for a three-person crew (pilot, co-pilot, and the on-board technician); (b) a wide screen located under the cabin; (c) TV camera and a model of the runway; and (d) instructor-pilot's panel enabling the recording of the course of the flight and the flight parameters. The pilot's activities related to take-off and climb to up to 200 m as well as descent from 200 m till the end of the landing maneuver shall be reflected on a screen in front of the pilot, in the pilot's field of vision. This means that each maneuver performed by the pilot appears on the screen in the form of changes in the environment, such as those during a real flight (e.g., tilting to the side causes a corresponding change in the ground plane on the screen). This simulator enables airplane crews to exercise all the elements of a flight typical of a passenger aircraft, such as take-off, climbing, level flight using radar equipment, positioning to landing, landing. The analyzed crew flights were performed in such a way that first the pilot sitting on the left controls the plane, and then the same flight was performed by the second pilot. During the flight, deviations from the prescribed altitude and speed parameters were recorded. The measure of the quality of the operator's activities was the percentage factor of the deviation from the standard in various flight phases. Twenty-four pilots aged 30–43 were tested. Among other things, a high correlation was found between the results of tests using the simulator and the quality of piloting in real flight in the two most difficult flight phases, i.e., take-off and landing.

Similar results were obtained by J. Maciejczyk, W. Kuzak, and F Skibniewski (1996) on a Polish flight simulator JAPETUS. Eighty students of the first year of the aviation school in Dęblin with the average age of 19 years were tested on this simulator. They performed seven exercises with a growing level of difficulty of the aviation task. There was a high correlation between the results of exercises on the simulator and the improvement of visual-motor coordination in the take-off and landing phase (r= 0.80).

The most difficult element of simulators is the visualization of external flight conditions, which depends on the graphical capabilities of computers, and which plays a huge role in the formation of aviation habits useful for working in real flight conditions. Modern generation of computers with a high level of data integration basically solved this problem, especially in the area of modeling and simulating control processes of the entire operator–technical object–environment system (Cacciabue 1999).

To sum up the benefits and limitations of simulators for training operator's efficiency, it is worth referring to the research by L. Żakowska, O. Carsten, and H. Jamson (2005), carried out using the SWOT method, which analyzes the strengths, weaknesses, opportunities, and threats of integrated simulator systems, demonstrating their value for practical applications. These studies show, among other things, that the so-called strengths of vehicle simulators include: 1) repeatability of each simulation, 2) parameters of environmental, traffic, and external factors geometry under absolute control, 3) possibility of simulation of each situation and event, 4) measurability of movement parameters with high resolution (speed, acceleration), and 5) operator behavior monitored and recorded in real time. Whereas, *weakness* concerns: 1) the relativity of the output, 2) the lack of standard protocols, and 3) the large number of output data to be processed. To sum up the advantages and disadvantages of simulators, the *opportunities* should be emphasized, including above all: 1) multidisciplinarity, 2) standards for evaluation of results, and 3) reliability of diagnostic tools. However, regardless of the presented evaluation, the development of simulation techniques related to the creation of dynamic visualization of the work situation closer to reality and the possibility of developing new software, taking into account visualization similar to the real conditions and the possibility of registering operator behavior and their analysis online, has been a fact for many years.

On the basis of the presented research and literature, despite the differences between the situation of the operator in the flight simulator and in the real flight, it can be stated that the advantages of training on simulators are unquestionable (Roessingh, 2005), and their development reflects the progress of technical civilization occurring at an ever-increasing rate (Koonce, 2002; Ogilvie, 2007; Vincenzi et al., 2009). An example of this thesis is the dynamic development of aviation ergonomics over the last half-century, as illustrated by Fig. 16.

Fig. 16 presents the consequences of the development of modern jet aviation triggered by the use of constantly developing newer generations of microsensors and microprocessors in the construction of avionics. As the analysis of this

Training of operator's action 117

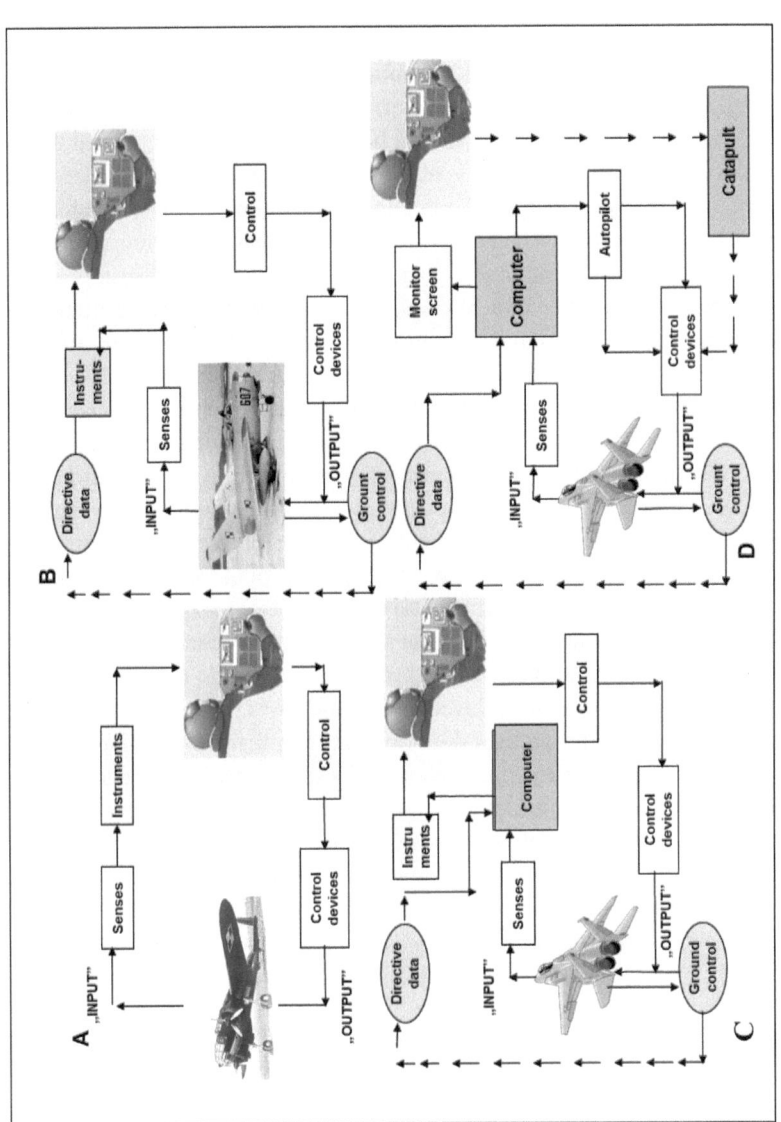

Fig. 16: Four stages of development of avionics in the 20th century, *significantly changing the aircraft operator's work* (own study)

figure indicates, the classic pilot-airplane system included manual control of the machine by the operator (Fig. 16A) using senses and motor systems. Generation of indications of pilot-navigation devices was carried out at a mechanical level. It was possible in the situation of such aircraft construction in the 1920s or 1930s, which in the control process provided information about the condition of the aircraft by means of several basic indicators (speed, attitude indicator, tachometer, fuel gauge, and altimeter). It should be noted that the flight speed oscillated between 100 and 200 km/h. With the increase in flight speed, there was a shortage of time needed for the pilot to process pilot-navigation information, which is why it was necessary to replace the existing aviation indicators with directive indicators (Fig. 16B), suggesting the direction of decisions to be taken by the pilot. Another radical change of the aircraft engine construction for jet engines in the 1950s and closing the gap between flight speed and the sound barrier, on the one hand, increased the number of indicators to several dozen, and, on the other hand, reduced the time needed to observe navigational equipment (inside the cockpit) and observations of the battlefield (outside the aircraft), to such an extent that it was necessary to introduce an on-board computer (Fig. 16C) as a means of supporting control and the decision-making processes. The next stage of development of modern jet aircrafts, breaking the sound barrier several times, practically eliminated the pilot from operator activities by introducing a computer-controlled autopilot (Fig. 16D), leaving it free to control air combat. Simultaneously, observation of the airspace is supported by communication with ground control and radar equipment. At the same time, an on-board ejection seat was introduced, which increases the feeling of safety in life-threatening situations. Further development of military and civil aviation is related to the use of satellite communications for observation of airspace, which will enable simultaneous contact with many ground control centers and flying objects, significantly shortening the time needed for human communication and increasing flight safety. If one considers the introduction of vertical take-off planes (like a space shuttle), which will break the sound barrier by several times, and will allow to cross the distance between New York and Sydney in a ballistic flight within 3–4 hours, it is possible thanks to the use of the latest microprocessor technologies (*hi-technology*). This requires high processing power online computers and specialized software, which generates increasing costs and development of flight training systems for operators, instructors, and maintenance personnel. This also makes it necessary for work psychologists to develop new paradigms for the selection of candidates for future operators of technical equipment. For example, by accepting the suitability of simulators for flight training, flight practice developed a system of continuous pilot training which, for each new aviation task, provides for a large repertoire of aviation exercises varying in terms of difficulty. The flight regulations assume that the appropriate pilot class is equivalent to the actual level of mastery of operator habits and aviation skills for a given type of aircraft, task, and conditions in which a real flight is to take place.

Many tests on pilots concerning their subjective evaluation of the application of flight habits learned in flight simulators to real flights indicate that this evaluation is not entirely positive, although it is much better in case of older, experienced pilots (e.g., Lempereur, Lauri, 2006), especially when it comes to the exercise of so-called special emergency cases (e.g., self-rescue maneuver of a helicopter) (DeVoogt, Van Doorn, 2007) or training risky decisions in emergency situations (Li, Harris, 2008). The amount of positive evaluation of flight simulators increased with the introduction of objective *online* registration of both flight parameters and pilot performance in modern commercial aviation simulators (e.g., Boeing 737N6 generation), which are very useful for discussing errors and improving flight training. For example, C.M. Björklund, J. Alfredson, and S.W. Dekker (2006) monitoring the training on a new generation Boeing 737N6 simulator, additionally using ophthalmic recording of pilots' eye movements, found, among others, 49 % improvement in efficiency of the tasks performed after 12 training flights (521 hours of training).

At this stage of the development of aviation ergonomics, there is still no consensus on the type of simulators that should be produced. For example, supporters of comprehensive and universal simulators believe that modern airplanes have many features in common and that a universal operational structure can be developed. Supporters of functional simulators rather focus on the specificity of operation on different types of aircraft and different phases of flight (De Voge, Bass, 2007). The current practice tends to prefer functional and specialized simulators, which from the psychological standpoint is more justified. Regardless of the substantive discussions on the development of simulators for different groups of operators, the need for their production itself does not raise any doubts, which is undoubtedly related to the development of computer technology and software useful for solving difficult problems of visualization of both the parameters of the simulated equipment (e.g., aircraft, spacecraft, car, sea ship, power plant, etc.) and the parameters of the working environment (e.g., climate, visibility, obstacles, failures, etc.) (Harris, ed., 1999). The development of computer software also led to the moving away from static simulators towards dynamic ones (e.g., mobile platforms installed on hydraulic cylinders), reflecting real flight conditions (Salden, Paas, Van der Pal, 2006). The unquestionable didactic value of modern simulators comes down to the fact that one can train various types of failures and dangerous work situations in a virtual environment, which cannot be trained in real operating conditions without compromising safety (Jasiński, 2005). For example, master mariner train steering ships to enter the expected ports on given lines using simulators. This allows you to achieve two training goals: (1) to define the role of the specific operator in the reliability of the O–M system and (2) to increase individual experience in dealing with stress caused by failure of the technical equipment. The possibility of cooperation with an instructor training the operator on a simulator, whose competences and experience significantly increase the effectiveness of vocational training, is not without significance. The presence of an instructor is particularly useful in emergency situations. Although the instruction to act in such situations, learned during the theoretical training, imposes a specific sequence of actions on the operator (e.g., in aviation there is a document called "Special events during flight"), it should be remembered that the acquired habits refer to so-called normal situations, in which it is not necessary to change the learned structure of action. Whereas, in an emergency situation characterized by emotional stress, it is necessary to radically change the existing conceptual model of action into a new operational model (action plan), which undoubtedly requires a change in the existing structure of actions. In such cases, the instructor's greater experience may prove to be very helpful on an *ad hoc* basis, as future effectiveness in real-life situations depends on gaining individual experience with emergency action. This is confirmed by own research on the effectiveness of pilots depending on their individual aviation experience (Terelak, 1988), which shows, among other things, different characteristics of the pilot's effectiveness in emergency situations related to professional experience. Thus, highly experienced pilots respond to an emergency situation in 33 % of cases in a time up to 3 s and in 11 % of cases – above 20 s; pilots with little experience – in 5 % of cases – up to 3 s and as much as in 33 % of cases – over 20 s. It is characteristic that the first group after a special training on a plane simulator increase their reactivity to 65 % of cases in a time up to 3 s and decrease only to 4 % in time above 20 s.

When assessing the accuracy of the measurement of operator's efficiency in three situations: scientific laboratory, simulator, and real working conditions, the latter situation should be assumed as the most accurate (e.g., length of service) but too expensive and extending the time of diagnosis. Taking into account the latter two factors: economic and time, the optimal option is to use a simulator for both the conditions and the structure of the operator's tasks.

Summing up the considerations on the psychological bases of vocational training with the use of simulators, attention was paid to the main objectives of this type of operator training and its limitations in relation to the actual working conditions. The first conclusion to be drawn is the methodological awareness that operator training does not end with the development of sensory and motor habits, i.e., strictly algorithmic working activities, because the habit is a necessary stage of mastering operator's activities but is the least complicated psychologically. Therefore, its shaping does not cause any major difficulties at this stage of vocational training with the use of simple functional simulators. The second proposal concerns the more complex development of the ability to identify specific, usually unforeseeable events in the process of operator's work. This type of training should take into account the quite significant individual differences in the ability to detect the necessary signals about the condition of the technical device and the environment, as well as the ability to make an appropriate diagnosis on this basis and make decisions on how to proceed. Psychological screening methods in terms of specific abilities and personality traits of operators, which can be used by instructors to individualize vocational training, are useful for this purpose. The third proposal concerns the development of the ability to cope with professional stress through the acquisition of experience with simulated and real stress factors at work, which is the next subject of consideration on vocational training in real working conditions.

4.4 Adapting the operator to the real working conditions

The third stage of vocational training of operators is training in real-work situations, which results from a more general, already classic problem of psychology of work, namely person's adaptation to work. It depends, among other things, on different attitudes towards work. The following three main attitudes towards work are highlighted in the literature on the subject: (a) punitive attitude – a person perceives work as a physical, moral, or economic constraint; (b) instrumental attitude – a person treats work as a means of satisfying needs; (c) autotelic attitude – work is perceived as a value, an objective in itself, a source of self-fulfillment in professional work (Reynolds, 1997). There are a number of detailed concepts of defining with this problem, three of which are worth mentioning: (1) the concept of fitting for a profession as such, known in the literature as "P–J Theory" – *Personality–Job Theory* (Edwards, 1991); (2) the concept of fitting for the physical working environment called "P–E Theory" – *Personality–Environment Theory* (Caplan, 1987); and (3) the concept of fitting for the organizational factors of work, known as "P–O Theory" – *Personality–Organization Theory* (Kristof-Brown, Zimmerman, Johnson, 2005, 996). There are also attempts at juxtaposing the validity of both approaches, P–J versus P–O (e.g., Carless, 2005), which, however, do not question the importance of the complementarity of both concepts. It can be assumed that P–J, P–E, and P–O fitting is a specific form of employee adaptation training, which is interactive and systemic in nature, i.e., it does not result only from job characteristics or solely from characteristics of a person, but from the existence of relations between them. Therefore, job fit is an important issue, as compatibility between individual's characteristics and job characteristics affects not only people's attitudes towards work, but also their effectiveness (Cable and Edwards, 2004).

P–E trainings related to adaptation to various physical factors of work (e.g., high vs. low temperature, hypoxia, accelerations, weightlessness, etc.) shall be discussed on the example of predictive effectiveness assessment of some aviation simulators. Generally, this stage of adaptation takes a certain amount of time (so-called work experience, probational period) and is also includes theoretical training. The health and safety instructors (occupational health and safety) are the ones who are most often responsible for this training cycle, which ends with obtaining an appropriate certificate. Having the appropriate standards of tolerating various factors, they select and assess their trainees positively or negatively (Koradecka, ed. 1997). However, this does not mean that they will be fully adapted when it comes to real working conditions. This is due to a number of factors, including but not limited to: (1) Weakness of the method of extrapolation on the basis of a single factor, even if precisely investigated using simulators of working conditions (e.g., thermal chamber or decompression chamber or centrifuges, etc.) in relation to the actual working conditions in which all these physical factors of work occur comprehensively (Tinsley, 2000). There are still too few mobile laboratories (e.g., training combat aircraft, "driving education" cars, etc.) which would determine the tolerance of the combined factors of working conditions. (2) The passive position of the taught and trained operator in the conditions of theoretical training and simulator training compared to the task situation and responsibility (financial, moral, social) in real working conditions. (3) Lack of tools to quantify the impact of the initial period of adaptation (seniority) to the actual working conditions in a specific operator occupation and at the same time the lack of a standard list of occupations. Excessive divergence in the classification of occupations and the constantly appearing new professional groups lead to a situation that psychologists are not methodologically prepared to solve this important social problem.

The detailed discussion of the concept of P–O shall be omitted, as it is a problem of the psychology of organization and has its own rich literature (Dunnette, Hough, eds., 1990–1992). The focus shall be placed only on certain aspects of the subjective fitting of P–E and P–O, mainly on the example of the pilot profession, which due to the necessity of cooperation in the process of work with other people (e.g., aircraft crew, communication with ground flight control

Adapting the operator to the real working conditions 123

centers, etc.) and the priority role of work safety is most accurately described in the psychological literature. The different classifications of cinematography professions, which are difficult to accept, shall be omitted as well, due to the fact that the criteria for separating particular professions and professional groups are incomplete[38] (Widerszal-Bazyl, Cieślak, Derlicka, ed., 1998).

With respect to the Multidimensional P-O Fit Model by Amy L. Kristof (1996), two aspects can be distinguished: supplementary and complementary, which result from different processes of perceiving work. The main difference between the above fit types is the adopted definition of the working environment. In the case of a supplementary fitting, the working environment is described by the people who function directly in it, and in the case of a complementary fitting also by people who do not come into contact with it. Thus, supplementary fitting occurs when an individual supplements, enriches, or has characteristics similar to those of the working environment (e.g., culture, emotional climate, values, goals and norms, etc.). The most commonly used indicator for this type of adaptation was compliance between individual and environmental values, which is in line with common understanding (Westerman and Cyr, 2004). Complementary fitting occurs not only when an individual's characteristics form a complete whole with those of the working environment, but also when the individual contributes something significant to that environment and not only relies on

---

38 From the formal perspective, the types of road vehicle operators in Poland are defined in the Regulation of the Minister of Economy and Labor of December 8, 2004, on the classification of professions and specializations for the needs of the labor market and its scope of application (Journal of Laws, No. 265, item 2644) and concern the operation of the following motor vehicles, the operation of which requires an appropriate driving license category: (1) Motorcycles and mopeds; (2) Automobiles: quads, passenger cars, campervans, taxis, rally cars, vans, trucks; (3) Special vehicles: concrete mixers, tractors, delivery trucks, cranes, municipal, garbage trucks, tanker trucks, etc.; (4) Bus vehicles: city buses and intercity coaches, minibuses, trolleybuses, double-decker buses; (5) Low-speed vehicles: agricultural machinery, combine harvesters, tractors, garden equipment, construction machinery, forestry machinery, road machinery, mining machinery, excavators, loaders, etc.; and (6) Special and police vehicles, firetrucks, prisoners, driving education vehicles, military vehicles, postal service vehicles, transport service vehicles.

As it can be seen from the above regulation, different groups of drivers require different types of licenses and detailed testing methodologies for nonspecific and specific operator efficiency. This problem is discussed in more detail, including operators of other railway, maritime, air, and space vehicles in another study of mine (cf. Terelak, 2015).

the exchange of resources, mutual contributions, and profits (Kristof-Brown, Zimmerman, Johnson, 2005).

When considering the fitting of subjective expectations of P–E and P–O, the e xperience of training in real-world conditions for pilots of passenger aircraft are worth referring to. On the one hand, it concerns the development of the skills of using intelligent computer-aided aircraft control, and on the other hand – the ability to manage the crew of a passenger aircraft after learning the CRM (*Cockpit/Crew Resource Management*) and ORM (*Operational Risk Management*) principles (Sherman, 2003). The scope of CRM training includes, inter alia, combining the knowledge, skills, and attitudes of those involved in a common task and increasing crew awareness of the role of work and team responsibility. This knowledge should translate into practical operational skills, manifested in the following six learned behaviors adequate for each aircraft type: (1) situational awareness (targeting attention and perception supported by computer-analyzed data on the condition of the technical device and the external environment); (2) task management (coordination of operations); (3) decision making (risk assessment skills); (4) task analysis (analysis of the flight phases); (5) communication (coordination of information exchange between crew members); and (6) crew coordination (ability to subordinate to crew management) (Salas, Cannon-Bowers, 2000). Initially, CRM consisted in increasing the effectiveness of communication between people operating in the airplane cockpit, and later the legitimacy of creating a team in order to increase responsibility for risky actions taken (ORM) was emphasized, which also result from the ability of the crew to cooperate with people outside the aircraft (e.g., air traffic controllers, etc.). The training concerns both decision-making processes, especially in the area of risky decisions as part of task planning, as well as in the task performance itself (Pedersen, 1999). Among other things, the following rules are taught to the crew members: (1) do not accept unnecessary risk, (2) accept risk when benefits outweigh costs, (3) make risky decisions at the appropriate level, and (4) anticipate and manage risk through planning. The introduction of CRM and ORM training is the result of the analysis of participation in air accidents and incidents "crew errors" caused by faulty communication between its members and "human factor" caused by incompetent communication between the aircraft crew and external (ground) agencies responsible for flight safety (Truszczyński, 2002). The problem of operator errors in the training process and real action is an introduction to a new theoretical and practical field, namely the problem of workload and reliability of people in the O–M system.

5. Reliability of the operator in the operator–machine system

The issue of operator reliability is borrowed from the technical sciences (electronics and cybernetics), where it is known as the safety of technical systems. Safety is generally understood as the reliability of the system, which is defined as the faultless operation of a technical device, fulfilling certain functions, under strictly defined operating conditions and in a defined period of time (Park, 1987). The indicator of technical reliability understood in this way may be, e.g., the average time until the first failure occurs. It is obvious that even under optimal operating conditions of a technical device, the reliability factor is always greater than zero and less than one. Therefore, there are practically no technical devices offering 100 % uninterrupted reliability, especially when operating in complex systems in which one of the elements is a human being.

5.1 Reliability as a technical category

The most well-known theories of operator reliability have their chronology, starting with stochastic theories (1900–1960), through causal theories (1940–1980) and systemic theories (1970–2014), to behavioral theories (1990–2014) (cf. Jamroz, 2008). In individual theories, both the emphasis on the mechanism explaining traffic incidents and accidents and the methodology of research vary. Thus, in *stochastic theories* (statistical theories), it is assumed that accidents are random, in line with the Poisson distribution model. This randomness excludes the forecasting of road accidents based on normal distribution of phenomena and is beyond rational human control. The causal theories of accidents include two currents: deterministic, referring to the current paradigm of the aftermath of events, and probabilistic, describing a set of separable factors responsible for operator's safety. An example of deterministic approach to events is the classical theory developed by Heinrich in 1931, known in the literature of the subject as "Domino Theory" (Heinrich, 1959), which assumes an influence on the current state of safety of the so-called antecedents, regardless of their nature. Systemic theories include the Edwards Theory (1981) SHEL model (from the first letters of the words: Software, Hardware, Environment, Liveware), which assumes interactions between the separate components of the model. The S component (Software) – covers the general principles of the operator's action in a specific working situation; the H component (Hardware) – concerns the ergonomic solution and technological capabilities of the machine; the E component (Environment) – covers the specific working environment; and the L component (Liveware) – concerns the psychophysical conditionings of operators (somatotype, height, sex, age, etc.) and personality (temperament, personality traits, general and special abilities, motivation, etc.), as well as health (e.g., fatigue, health). *System theories* are a step in the right direction, as they exclude the deterministic impact of one element, although their accuracy is more difficult to verify, especially in case of operator safety forecasts, because they require multifactor data analysis, as it was illustrated by W. Haddon's (1980) matrix, which is well known in the literature and is considered a precursor of an interdisciplinary approach to operator safety description. *Behavioral theories* emerged from methodological difficulties in the holistic analysis of machine operator's safety; they were pioneered by G.J.S. Wilde (1994), who focused on assessing the "desired level of risk" in a given society and at a given time. Psychological versions of behavioral theories are referred to, among others, by the operator's safety model by J. Reason (1990), known in the literature as the "Swiss cheese model". This model assumes that incorrect operation of the operator at any time is preceded by broadly understood antecedents (predecessors). In his error model, GEMS operator (Generic Error Modeling System) J. Reason (1997) distinguishes between "operator errors" and "human factor", because they have a different psychological interpretation. Thus, the very analysis of the effects of the operator's errors provides a basis for distinguishing a typical operator's mistake and the participation of other people, as well as external circumstances conducive to errors.

Quantitative measurement of causes of operator errors, in the opinion of D.A. Wiegmann and S.A. Schappell (2003), is possible with the help of a tool constructed by them, known in the literature of the subject as: The Human Factor Analysis and Classification System (HFACS), which allows for the randomization of human factor contribution to operator safety at four error levels: (1) hazardous activity – active factor, related to operator's errors; (2) conditions conducive to hazardous activity; (3) inadequate supervision; and (4) organizational factor. First type errors, being the most frequently made, testify to bad *situational awareness*.

Summing up the review of groups of theories explaining the occupational safety of operators of various technical devices, it should be stated that currently no single general model dominates, although there are still partial models referring to the previous ones, namely: models of system reliability, disaster models, awareness models, vigilance models, etc., which differ in the level and sophistication of quantitative methods of analysis. Pragmatism indicates the possibility of using several complementary models, e.g., statistical, causal, and behavioral, suggesting the possibility of developing metamodels of operator's safety, an important element of which is operational reliability.

5.2 Reliability as a psychological category

The concept of reliability, transferred to medical and social sciences, is usually understood as an individually changing human characteristic that stabilizes the results of a person's work under strictly defined conditions of workload (Williams, 1958; Niebylicyn, 1969). The introduction of the concept of human reliability to work psychology is a very distant and imperfect analogy. Of course, this concept has an operational and descriptive status, not a theoretical one, despite the existence of several of its models (e.g., Franus, 1978; Ratajczak, 1988).

In the first sense, reliability is understood in physiological and psychological terms as the efficiency of faultless human functioning, defined as a highly complex self-adjusting system. In this sense, an indicator of human reliability may be, e.g., the overall adaptability to specific environmental conditions at a given time or the effectiveness of the functioning of individual mechanisms of personality control. Such an understanding of reliability is expressed in Polish psychology, e.g., regulatory theory of personality (Reykowski, 1966) or temperament (Strelau, 2006).

In the second sense, the human reliability boils down to the efficient operation of person as an element in the O–M system. A quantitative indicator of such an understanding of human reliability may be the probability of effective and timely completion of an action and its results (cf. Łomow, Płatonow, 1984). Psychologists, often referring to statistics of accidents and catastrophes (railway, automobile, aviation, etc.) suggest that in the O–M system people are more unreliable than technical objects. This statement is only partially true, and only in this part, when we define human reliability by analogy with the technical reliability of a machine. Often when analyzing an actual person in their operator activity at the level of sensory cognition, it can be assumed that technical devices can successfully replace some of our senses at the stage of signal (information) detection, especially if we are restricted by a time limit, sometimes exceeding the time of a psychomotor response. However, if we refer to the regulatory function of human intelligence, which is responsible for "intelligent" behavior, especially in new and unexpected situations, the reason behind ergonomists' opinion that a person in the O–Machine system should be treated as a guarantor of the reliability of the entire system is quite understandable (O'Hare, 2006). So far, the percentage ration of the human factor in the reliability of the entire O–M system has not been established, even with the use of the neurosensory Theory of Functional Systems by PK Anokhin (1970). One of the reasons for this is the significant differences between the two components of the O–M system. There is a difference in the functioning of the machine, which is of a step nature, which means that there are no intermediate states between the states of functioning and immobility, and in the functioning of a person, whose physiological and psychological processes are continuous; and the transition states (e.g., loss of consciousness, fainting, sleepiness, inattention, etc.) are the causes of errors in action. In addition, it is not without significance to understand the differences that the functioning of the machine is linear in nature, which means that the output response is the sum of the responses to individual signals, while in humans the final response is not the resultant response to all the information received at the input. Other differences are also highlighted in the literature on the subject, such as the work pace, rhythmicity, susceptibility to fatigue related to the daily rhythm, motivation to act, etc. The last significant difference can be reduced to the logic behind the machine's functioning (algorithmic program) and that of the operator (heuristic program, which often allows to operate irrationally, incompatible with the applicable procedures). These differences between the machine and people are the basis for rejection of the common definition of reliability of the entire O–M[39] system. There are examples of attempts to quantify the reliability of the O–M system, according to which the reliability of this system is the probability that a given system will perform its functions satisfactorily in a given time and conditions. Indicators of reliability understood in this way include: (a) the readiness of an operator expressed in terms of a specified probability of engaging in an activity at any time, (b) the average work time between two operator errors, (c) the sum of errors within a given interval, (d) the percentage of fault-free tasks performed, and (e) the probability of fault-free operation for a specified interval (Łomow, ed., 1977; Nagel, 1988). Currently, however, the quantitative analysis of the reliability of the operator's work, which inventories only errors and omissions as well as their percentage distribution in time (descriptive approach), has been abandoned in favor of qualitative analysis, which explains the nature, importance, and psychological mechanism of errors in the process of operator's work in specific conditions (explanatory approach). An example is stress psychology, which tries to explain the causes and effects of passing

---

39 Such attempts were often made in the past, e.g., in cybernetics, where human reliability was defined as the reliability of highly specialized computer-controlled machines, as the ability of a human being to carry out the operator's tasks entrusted to them without error under certain conditions and in a certain period of time (Bobniewa, 1969).

through the process of working with technical equipment from the "load zone" (*Techno-Eustress*) to overloading (*Techno-Distress*) (Tarafdar, Cooper, Stich, 2019). P.A. Hancock and P.A. Desmond (2001) devoted a special monograph to this issue entitled *Stress, workload, and fatigue*, inviting the most prominent representatives of psychology, physiology, and ergonomics to cooperation. The monograph presents many interesting theoretical models and proposals for workload testing methods. Noteworthy is, among other things, the double workload model, which is a function of: (1) the technical equipment and the task being performed; (2) characteristics of an employee's skills and professional experience. This model is referred to by M. Fąfrowicz and T. Marek (1999), who distinguish three similar notions that can be used to describe workload in its different scopes, i.e., mental stress, fatigue, and mental load. To this end, they distinguished two types of effort. The first type is the mental effort related to the level of difficulty of the task performed (mental strain), and the second type is the compensatory effort, related to the control of body states (mental stress and mental fatigue). The authors refer to the normative definition of stress proposed by the International Organization for Standardization (ISO 10075, 1991), which on the one hand points to situational factors causing mental stress (requirements imposed by the task performed, physical and social working conditions, and social factors outside work), and on the other hand to individual characteristics of a person (such as: aspirations, system of attitudes, skills, abilities, knowledge, experience, health condition, age, etc.). The final effect of a mental state depends on the relationship between the elements of the working environment and the specific characteristics of the individual. This transaction is important for determining the magnitude of workload.

Other attempts to index the workload of operators of highly specialized technical objects, such as high-maneuver airplanes, spacecraft, etc., can also be found in the literature. An example of which is the *NASA Task Load Index - NASA-TLX*, which covers six load aspects, namely: (1) mental demand (type and level of mental activity), (2) physical demand (type and level of effort), (3) temporal demand (required time and pace of work), (4) overall performance (success rate and level of satisfaction), (5) effort (consumption of physical and mental energy), and, (6) frustration level (stress symptoms). All these aspects of workload can be assessed using special 20-degree scales (1 for low load and 20 for very high load) (Hart, Staveland, 1988; Wilson, Corlett, 1995).

As in the case of theoretical difficulties, there are methodological problems related to load and fatigue testing. In the literature on the subject, various workload model proposals can be found, including the model of interaction of environmental and personal resources by Van Vegchel, De Jonge, and Landsbergis

(2005), which includes the proposal of three complementary methods of workload testing, namely: (1) measurements of effectiveness under real-world conditions or task samples (e.g., work errors, extended operating time, etc.) (Schellekens, Maijman, 1999); (2) psychophysiological measurements (e.g., psychophysiological cost as a response to workload or workload estimated by means of heart rate measurement, ECG, EEG, etc.) (Kaber et al., 2007); and (3) subjective measurements of workloads using specially constructed assessment scales. It is worth noting the subjective methods of measurement, as they are available to all psychologists, without the need for special equipment. An example of this can be the concepts known from the literature of the subject as: *Subjective Workload Assessment Technique – SWAT* and *Workload Profile – WP* (Biernacki, Bicka-Capała, Tarnowski, 2007). This SWAT method allows to characterize three subjective aspects of workload: *time load*, *mental effort load*, and *stress load* and their reference to three levels of incidence. This gives a combination of 27 subjective assessments of workload, which in turn need to be carried out by the examined person in ascending order and assessed from the point of view of proportion of individual load dimensions (Meshkati, Hancack, Rahimi, 1995).

The advantage of the presented behavioral, physiological, and psychological methods of studying workload is, on the one hand, their complementarity and, on the other hand, the possibility not only to describe the types of workload, but also to study the relations between them, i.e., the possibilities of one type of workload influencing other (Averty et al., 2004). A separate problem is online telemetric monitoring of physiological and psychological parameters of fatigue at the operator's workplace in ground, aviation, and cosmic conditions (Truszczyński, Nowicki, Achimowicz, 2012). It is also important that the recorded psychophysical parameters in hazardous conditions are coupled with automatic systems, which can adjust to the state of the operator and modify the modes of action outside the state of consciousness (e.g., at high accelerations +Gz sensors placed on the pilot's baroreceptor anticipate the loss of consciousness by activating the automatic pilot, which leads the aircraft out of the life threat zone) (Pope, Bogart, Bartolome, 1995; Stephens, Pope, 2014).

The two concepts of O–M system reliability presented above, quantitative and qualitative, are acceptable in general terms, but within them it is not possible to solve specific issues related to occupational safety. There is no consensus on a uniform theoretical position on human reliability in the O–M system and no unambiguous answer to the question: Does ensuring optimal workload conditions, as suggested by proponents of concepts based on the known theories of Optimum Activation by D.O. Hebb (1965) and Optimum Stimulation by C. Leuba (1965), really provide a certain guarantee of reliable human activity at work? However, at the research level, the undertaken issue of reliability of the O–M system contributed to intensification of research on the influence of various factors of the working environment on human reliability conducted from a theoretical ergonomic (Franus, 1977) and work psychology (Ratajczak, 1991) as well as the perspectives of methodological scaling of workload (Dudek, Markowska, 1986). These studies clearly indicate that the practical consequences of the operator's actions in ranges exceeding optimal conditions, pushing them in extreme directions, are always associated with the risk of errors.

Summarizing the current considerations on reliability as a psychological category, it should be stated, among other things, that solving the problems of reliability of the O–M system exceeds the capabilities of one discipline. The existence of the rich literature on some of the determinants of reliability of the O–M system resembles a situation that arose at one time in ecology, where almost unlimited number of coefficients of correlation between environmental factors and characteristics of living organisms were obtained (Pearson, 1999). This led at one time to the misconception that the environment is such a compact entity that it should not be dissected into individual factors. The first of these views, despite the quantitative approach, led to a multitude of controversial research results, while the second, holistic view, entirely qualitative, is more useful for nonscientific deliberations or for making an individual diagnosis of a particular person than anything else. A way out of this situation is leaning towards increasingly interdisciplinary research, which in the field of reliability of the O–M system suggests a systemic approach, which does cause new methodological difficulties. However, this does not mean that within individual scientific disciplines there are no attempts to solve partial problems, among which occupational safety issues were of great importance for many years (Rasmussen, 1986). There are three main concepts of occupational safety: psychological, sociological, and systemic.

5.3 Psychological concepts of occupational safety

Since there are too many unsolved theoretical and methodological problems with regard to the reliability of the O–M system, it is worth paying attention to the issue of occupational safety from the perspective of ergonomics, occupational medicine, and psychology (Dahlberg, 2001). Well-known English theoretician and occupational safety researcher W.T. Singleton (1979) believes, among other things, that occupation safety is understood in negative rather than positive terms, because primarily its effects are mentioned rather than psychoprevention. This is confirmed by world statistics on the causes of aviation accidents, which include the following categories: human factor – 68.8 %, aircraft – 14.2 %,

132 Reliability of the operator in the O–M system

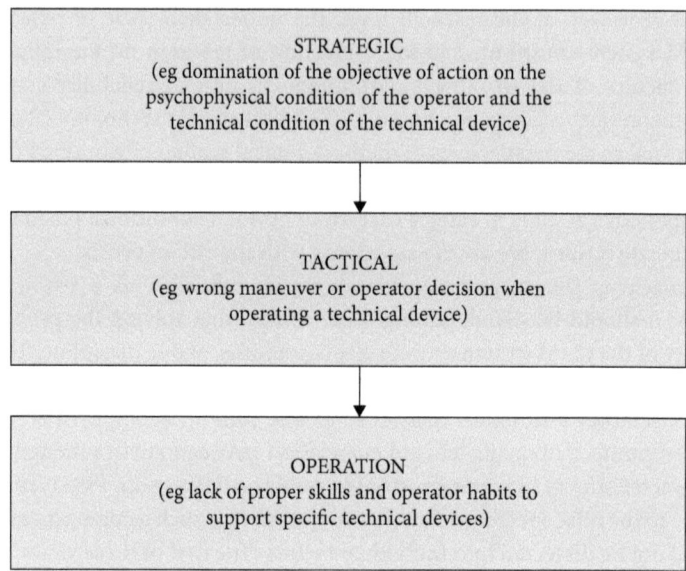

Fig. 17: Three levels of error of operators of technical devices (own elaboration)

weather – 6.7 %, ground air traffic control – 4.3 %, maintenance – 3.4 %, others (e.g., birds) – 2.4 % (Tsang, Vidulich, 2003). Therefore, it is clear that the causes of aviation accidents are dominated by the so-called human factor, which also includes not only the operator's error, but also other crew members and ground handling staff. However, a thorough analysis of the individual's participation in an accident encounters serious difficulties in perceiving such multidimensional phenomena that accidents are. A "dangerous operation" of the operator causing an accident, similarly to the "iceberg" model, is available to direct observation only in the 1/7 ratio, which is why it requires a detailed analysis of the so-called antecedents and nutrores explaining the causes. The remaining part concerning indirect causes of the accident, i.e., the so-called antecedents, is very complex and difficult to investigate from the methodological standpoint (Reason, 1990). It is not without significance for the classification and analysis of operator's error to define an error at least at three distinctly different levels: strategic, tactical, and operational, as shown in Fig. 17.

As Fig. 17 shows, each of the three levels of defined error requires separate expertise of the operator and the specificity of the technical device. These problems are dealt with in detail in error theories attributed either personally to the operators themselves or to the so-called human factor or, finally, both approaches together.

Studies on drivers suggest that most experienced drivers control vehicles automatically, quickly, effortlessly, and flawlessly. Automation may occur at different levels of control of the situation on the road and depends, inter alia, on factors such as: knowledge (theoretical driver training), knowledge of the rules of operation (highway code), and skills (practical training). The latter factor comprises three levels of automatic driver behavior: strategic automation, tactical automation (maneuvering), and operational automation (controlled). *Strategic automation* is characterized by, e.g., use of navigation in unknown terrain (knowledge), selection of known routes (rules of operation), and use of known routes to shorten the route (skills). The tactical strategy uses knowledge (e.g. regarding slip control), knowledge of the rules (e.g. when overtaking other veihcles on the road) and skills (e.g. overcoming difficult road curves). . *Operational automation* is characterized by (controlled) concern, e.g., drivers starting the first ride (knowledge), driving in an unknown vehicle (rules of operation), and using vehicles in everyday communication (skills). Distinguishing stages in the above model: strategy, maneuvers, and operational activities in the above model is primarily of illustrative importance, as, e.g., the strategic stage includes such elements of the driver's activity as, e.g., route planning and choice and time of driving. Tactical decisions to maneuver a vehicle are characterized by a certain driving style. In the operational phase, decisions are considered in specific traffic situations, e.g., relating to speed control, braking, and signaling. Thus, operator's abilities requires complex action and should be taken into account in the judgment process from a road safety perspective, as, e.g., the driver's strategic skills in anticipating braking in relation to the assessment of the distance of a road vehicle from an obstacle and its speed are the basis for a strategic decision, which may result in an accident or its avoidance (Rock & Harris, 2006).

The model of James Reason (1990) is a useful model for analyzing the causes of human factor–related accidents and is known in the literature as the "Swiss Cheese Model". An analysis of the causes of accidents in this model indicates that the operator's error is a consequence of previously occurring errors of the operator, the condition of the machine, work organization, etc., which had not previously been noticed and eliminated. In his *Generic Error Modelling System – GEMJS*, Reason (1997) distinguishes between "operator errors" and "human factor" because they have a different psychological interpretation. Psychologists involved in explaining the reasons for an error made by a pilot-operator must take into account all the relationships between the pilot and the machine, crew, other people, procedures, etc. These relationships may provoke circumstances that are conducive to, or even provoking errors in the operator's actions. Therefore, the analysis of the "human factor" component of the aviation accident investigation methodology, approved by the American National Security Agency and the Federal Aviation Administration, covers the following aspects: (1) behavioral (analysis of 24–72 hours of life history before the accident, operative action, pilot habits, and important life events); (2) health (general health condition, sensory reactivity, tendency to addictions, and fatigue); (3) operational (training, flight experience, operating procedures, and command tactics); (4) task (communication, task complexity, time pressure, and workload); (5) equipment (cockpit, status of indicators at the moment of an accident, recording of objective flight control from a "black box", and position and condition of the aircraft seat); and (6) environmental (weather conditions, internal conditions, lighting, noise, vibration, and overloading).

For example, K. Gerbert and R. Kemmler (1985) carried out an analysis of the causes of 1,448 aviation accidents according to the above model and, after subjecting them to factor analysis, developed the Teppe-Haakonson Pilot Reliability Model (Haakonson, 1980). According to this model, there are four main causes of aviation accidents, namely: (1) current operational capability of the operator (flight execution); (2) dominant "external" factors determining the occurrence of an error and critical situation; (3) dominant "internal" factors determining the occurrence of an error and critical situation; and (4) pilot action requirements that exceed current operator abilities, resulting in errors, creating an "accident zone". As part of the external factors, the following component factors were identified, e.g: (a) environmental (e.g., rapid transition from automatic to manual flight regimes, limited visibility due to weather, turbulence, vibrations, accelerations, terrain elevations, etc.); (b) task-related (low altitude flight, formation flight, tactical exercises, orientation flight, air combat, aerobatics, approach to landing on an alternate airport, etc.); (c) organizational and supervision-related (pre-flight time pressure, sudden changes in flight schedule, inadequate pre-flight checks, supervision pressure, poor crew coordination, faulty communication with the flight controller, etc.); and (d) technical and structural (failure of technical subsystems, inadequate cabin design, deficiencies and defects in pilot's technical equipment, etc.). In the case of internal factors, attention was paid to its components such as: (a) somatic (poor current efficiency of the pilot's body, fatigue, etc.); (b) psychological (stress before treatment, excessive motivation for achievement, high level of emotions in the course of performing tasks, lack of awareness of risk or too high self-confidence, etc.); and (c) aptitude (little aerial experience, insufficient pre-flight training, lack of knowledge of the document "special events during flight", lack of experience with flight stress situations, etc.). This model is useful to describe four categories of pilot in-flight errors, namely: "input" – attention and perception errors, and "output" – decision-making and sensory errors. These four categories of errors define both the operator's reliability and operational safety. Detailed operator reliability models such as the Teppe-Haakonson model presented above are, according to W.T. Singleton (1979), a basis for distinguishing three aspects of optimization strategy of operator's work safety, namely: (1) *Operator safety* – taking into account the narrowly understood interactions within the O–M system, which consists of two psychoprophylactic strategies: (a) minimizing the probability of operator error through appropriate psychological selection and vocational training; (b) maximizing the probability of operator error correction by, e.g., monitoring the operation of technical components and installing alarm devices, signaling the failure of technical equipment, etc.; (2) *Security as an emotional climate* – which depends, among other things, on organizational culture and work ethos, which should be preserved by means of propaganda (e.g., posters, training films, etc.); and (3) *System security* – understood as an effect of an optimal division of functions between the operator and the machine (Ek, Akselsson, 2007).

5.4 Systematic concepts of occupational safety

The classic system model of occupational safety known in the literature as the SHEL model is a transaction of its components such as: Software, Hardware, Environment, Liveware (Edwards, 1979). *Software* – refers to operator's action procedures and work regulations; *Hardware* – refers to the ergonomic solution and technological capabilities of the machine; *Environment*– covers both physical and social environment; *Liveware* – refers to the operators themselves with their constitutional equipment and personality as well as health.

On the one hand, the systemic model can form the basis for the interpretation of errors in operator's actions, and on the other hand, it can be helpful in explaining, controlling, and forecasting occupational safety. However, it should be added that these projects are extremely complicated, because the system approach uses a specific understanding of the "system feature", defined as a property of an element of the system, which does not result from it in a natural way (directly), but through the integration properties of the whole system. For example, in a road vehicle driver's safety system, the operators are at the same time characterized not only by properties specific to them, but also by a new nonspecific system characteristic resulting from driving a specific type of car

136　　　　Reliability of the operator in the O–M system

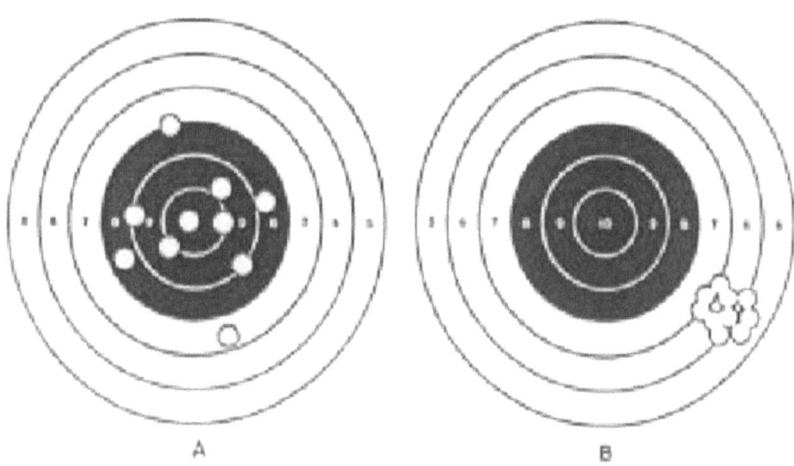

A　　　　　　　　　　　　　　　B

Fig. 18: The model of operator errors: stochastic (A) and systematic (b) (own elaboration)

(passenger car, truck, racing, ambulance, etc.) and performing a specific task. Thus, the systemic approach excludes the analysis of each element of the system separately with a preference to place emphasis on specific interactions between all components of the system. Such an approach requires interdisciplinary solutions, as it exceeds possibilities of one discipline, methodological possibilities in particular. In addition, this also applies to the absence of a uniform quantitative criterion on the basis of which it can be clearly established whether an emerging error is systemic or incidental. A proposal for such a quantitative criterion is presented in Fig. 18.

As shown in Fig. 18, a stochastic error is not of a systemic nature and is extremely difficult to predict. For these reasons, among others, the commissions for investigating transport accidents and catastrophes (land, air, and sea) are made up of a number of specialists (Terelak, Szczechura, 1998).

Knowledge of the systemic occupational safety model also has practical implications for general principles as well as specific ways of optimizing reliability and occupational safety. Outside purely cognitive reasons, accident investigation makes sense as long as it leads to increased occupational safety through prevention. Preventive measures should be applied already at the stage of finding "risky situations", which first have the status of "indicative of an accident" before it actually happens (Duffey, Saull, 2008). Five groups of methods for optimizing occupational safety can be formulated: (1) optimization of ergonomic solutions for the workplace (e.g., pilot's cockpit, etc.); (2) optimization of vocational training programs and specialist training for operators; (3) optimization of employee motivation (e.g., interests, pay, job satisfaction); (4) optimization of communication procedures between the operator and the manager, and possibly other team members (in case of teamwork) as well as social support at work; and (5) optimization of selection procedures and criteria as well as medical and psychological selection. An example of optimization of operators' occupational safety can be the criteria for analysis and classification of errors of operators operating in *US Navy/Marine Corps, US Army, US Air Force, US Coastguard*, which are specified in a document entitled: *Human Factors and Classifications Analysis System – HFACS* (qtd. Wiegman & Shappell, 2003). The literature on the subject emphasizes that the analysis of operator's errors as a method of training in occupational safety contributes to a significant reduction in failures and accidents (Hobbs, Kanki, 2008).

5.5 Psychological discriminative selection as a method of increasing reliability of the operator-machine system

The idea of psychological discriminative selection of people in terms of their individual suitability to perform civic tasks dates back to ancient times, or more precisely to times of Plato (428–348 BC), who in his treatise entitled "Republic", concerning the organization of an ideal state, drew attention to the existence of individual differences in the abilities and dispositions of citizens, which are the reason for the diverse effectiveness of their functioning. This Platonic paradigm of individual differences became the basis for all contemporary concepts of professional selection (e.g., Breaugh, Starke, 2000; Meglino, Ravlin, DeNisi, 2000; Bracken, 2007).

Let us recall that the earliest systematic and standard method used was psychological selection of military equipment operators in the US Army during World War I, when the *Army General Classification Test (AGCT)* was introduced. It was mainly aimed at excluding army candidates with a low level of intelligence due to the safety of handling of weapons entrusted to them (Jones, 2007). The usefulness of the methods of psychological selection in the army was evaluated very positively, which was confirmed by a real bloom of psychological tests constructed by a team of American military psychologists under the direction of J.P. Guilford during World War II. They were intended mainly for the "fast-track" preparation of proper operators of aircrafts to the air force organized in the USA from scratch. The air force was to quickly transport the American army across the ocean to Europe (Carreta, 1993).

Psychological tests are now used worldwide to test operators of different types of technical equipment, from simple (e.g., lathes) to very complex (e.g., new generation aircrafts and spacecrafts). Due to the rich literature of the subject, the historical aspect of this issue shall be omitted (cf., e.g., Schmitt, Chan, 1998; Scherbaum, 2005).

Since there are no universal psychological tests that could be used to examine different types of operators of machines with different levels of complexity, each recruitment and psychological selection process must be preceded by an analysis of occupations and, within it, by an analysis of specific jobs (Brannick, Levine, Morgeson, 2007). However, attention should be paid to an important methodological issue related to the preparation of a battery of diagnostic tests, which should be preceded by an analysis of occupations and psychological characteristics of the operator's post.

5.5.1 Analysis of occupations

The analysis of occupations is a consequence of the fact that in each organizational structure that pursues different objectives, its members represent many specific occupations in which they pursue partial objectives. Thus, the analysis of occupations covers several levels, such as: (a) identification of the place of the profession in the organizational structure at the workplace, (b) detailed characterization of the operator's post, and (c) valuation of a specific post. The analysis of occupations is defined as a detailed description of the tasks that make up a given profession and the relationships that occur in relation to other occupations, as well as a description of the requirements to be met by a given occupation with regard to knowledge (education), abilities (intelligence and talents), and professional skills, which determine its effective use[40].

Apart from encoding information for demographic purposes, the aim of classifying occupations is to reflect the sociological structure of citizens in a given country, e.g., based on remuneration or social prestige and creating a universal ranking of social usefulness in social awareness (Domański, Sawiński, Słomczyński, 2007).

---

40 In various countries there are at least two classifications of occupations: governmental (e.g., in Poland "Classification of Occupations and Specializations" introduced by the Minister of Economy and Labor of December 8, 2004; Journal of Laws, No. 265, item 2644) and sociological (Sarapata, ed., 1965), or psychological (Bańka, Chirkowska-Smolak, 1994).

When starting to develop a procedure for acquiring human resources from outside a specific organization, company, workplace, etc. (recruitment procedure), or useful for changing one's professional position within the current organizational structure, or in case of reorganization (selection procedure), the *classification of occupations and posts* in a specific organization must be taken into account (Kowalewska, 1965). This is usually related to the inventory of all possible post existing in the organizational structure of an enterprise and the place they occupy in the hierarchy of importance, resulting from the role of the implementation of a specific part of the group task. Only the *post assessment* – determined by the minimum qualification requirements for a specific employee, who must perform precisely defined activities related to the implementation of group objectives – becomes the basis for the work psychologist to develop the so-called recruitment criteria and the selection of research methods. Work psychologists draw attention to significant differences resulting from the perception of the operator's occupation and organization itself, within which there are significant differences in the expectations of candidates for specific professions, resulting from differences in awareness of the future professional role: imagined, perceived, accepted, and exercised (Lauver, Kristof-Brown, 2001), especially if dealing with new professions that go beyond the old classification criteria (Rollag, 2007). This also calls into question the diagnostic accuracy of a single psychological selection procedure for a specific occupation and suggests a multistage procedure, especially as no universal psychometric tools for analysis of a post can be found in the literature on the subject (Biela et al., 1992).

5.5.2 Analysis of a post

Analysis of a post requires precise answers to the following questions: (a) Why was a specific post created? (b) What is its place in the organizational structure of the enterprise? (c) What results are expected from a particular post? (d) What means of action are available at the post? (e) What are the requirements for candidates for the specific post? (Zinczenko et al., 1969). The most important is the answer to this last question and the ability to predict the intellectual properties, skills, and personality traits of a candidate for a specific operator post, etc. It should be noted that there is an asymmetry between a post and the subjective expectation of a candidate for this post. For example, the large discrepancy between expectations and perception of the future job is a source of cognitive dissonance. On the other hand, the discrepancy between the expectation and acceptance of a post sometimes leads to the formation of unrealistic motivation, which is a serious premise for elimination already at the initial stage of a job interview. An example of such research on the discrepancies between the expected professional role and the accepted role can be own observations concerning hidden factors disqualifying or limiting the so-called aviation abilities, which were the basis for eliminating candidates with the so-called neurotic-compensatory motivation, which generally consisted in the lack of real interest in aviation and excessive concentration on secondary aspects of military aviation (e.g., a beautiful uniform with a dirk, impressing friends, control over a "flying machine", etc.) (Terelak, 1971). Thus, the analysis of occupations and posts aims to reduce this discrepancy through the selection of recruitment methods that are most likely to accurately assess the capacity of the operator at the level of the role performed according to its definition. This definition of a post takes into account the rationale, objectives, as well as the criteria for its description and evaluation in at least three aspects: (a) *Competences* – defined by requirements of a given post. The most common attributes of competences are theoretical knowledge and seniority, intelligence and special skills, as well as some personality traits (e.g., assertiveness, ability to coexist in a group, etc.). Interpersonal competence plays an important role in personnel selection processes for positions requiring group or managerial work; (b) *Work attitude* – including the employee's attitude towards the job, including, inter alia, motivation, reflectiveness, independence, creative initiatives (if required), etc.; and (c) *Scope of responsibility* – concerning the answer to the question of how a given post affects the company's global result and how a given employee performs on this post, also (if appropriate) what is the scope of independent decisions (with financial consequences)? (Zalewska, 2001)

5.5.3 Recruitment and psychological selection

The analysis of occupations and post is carried out using a number of qualitative and quantitative methods. Among qualitative methods, which are popular in psychology of work again, the following can be included: *linking data* (casual conversation, interview, free speech method, narration, etc.) (Stokes et al., 1994); observation of behavioral samples (observation sheets, photography of work, etc.); conference techniques (panel discussion with experts on working methods and achievements); and various types of diaries describing activity at a given post (Filding, Folding, 1986). The latter two quality techniques are most often used in the probational period (internship). Quantitative methods include psychometric tests of mental and psychomotor skills (specialist equipment) and personality tests (questionnaires, scales, etc.) (Scherbaum, 2005).

After the analysis of the occupation and the post, this knowledge needs to be translated into the assessment of skills, which is the basis for the recruitment of

Psychological discriminative selection 141

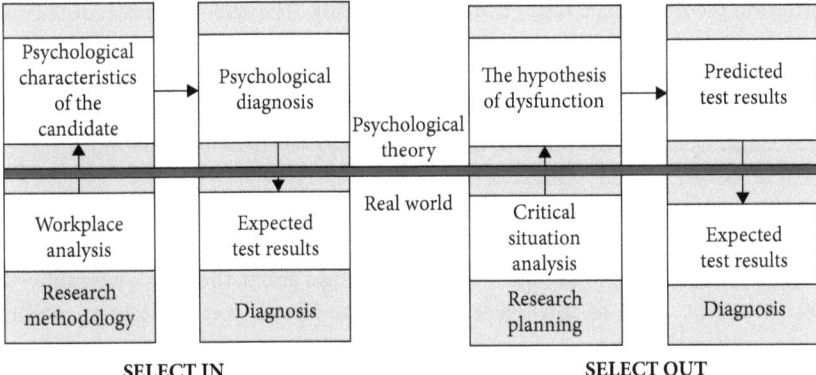

Fig. 19: Model of the structure of the psychological diagnosis in the situation of recruitment (select in) and secondary selection (select out) (own study)

candidates for a specific occupation, or psychological selection – related to the development of professional qualifications. The structure of both procedures is shown in Fig. 19.

As shown in Fig. 19, the procedure of psychological diagnosis in the working environment of an operator has two forms: (1) recruitment *(select in)* for the occupation, also called selection, includes procedures for collecting biographical and professional data (qualitative data) about the candidate who has not yet been known and the results of psychological tests, used to match the broad psychological profile of the candidate to the current professional profile developed earlier by the employer on the basis of the analysis of the occupation and a specific post; and (2) actual selection *(select out)*, concerning the assessment of suitability of persons already engaged in a specific occupation, and thus the already known traits – the so-called strengths and weaknesses (e.g., performance of duties at the post, level of adaptation to company requirements, type of dysfunctions and conflicts) in order to maintain the job, terminate the job, or give promotion (Hannum, 2008). The psychological selection procedure itself, which is a multistage procedure, is only a part of the recruitment system. The procedure consists of the following diagnostic steps: (1) initial interview, (2) document analysis, (3) medical examination, (4) psychological examination, (5) vocational (task) tests, and (6) probational period. At each of these stages, the candidate may receive a "positive forecast" or a "negative recommendation". The final evaluation of the procedure on a decision basis, although burdened with an averaging error of a qualitative criterion (e.g., character traits) with a quantitative criterion (score in scale tests) is final and binding. The selection procedure is of a rather positive nature, as it is often associated with the promotion or professional development of an already relatively familiar employee, which involves the need to raise qualifications, through the necessary professional training at this stage: operator (in relation to the machine) or social (in relation to other people) (Carretta, Ree, 2003; Carless, 2003).

However, the basic method of recruitment and psychological selection are professional tests, which are defined as specially constructed behavioral tests, which provide the psychologist with knowledge about the examined spheres of the psyche, such as intelligence (intelligence tests), special abilities (ability tests), personality tests (personality tests), knowledge (school tests), which meet the appropriate psychometric requirements of the so-called good test, i.e., they are: (a) standardized (they have the same test procedure), (b) normalized (i.e. conversation from raw to statistical result), (c) accurate (known accuracy with which a given test measures what it is supposed to measure according to theoretical assumptions or with the results of a similar test), and (d) reliable (having relative stability of time of the results obtained by the same tested person).

More than 100 years of experience with the use of tests for recruitment and selection studies proves that apart from theoretical reasons (e.g., models of professional requirements and human capabilities) and methodological reasons (e.g., development of psychometric and professional methods: personal pattern of the employee, characteristics of professional requirements, news tests, competence tests, etc.), the economic reasons for comparing the costs of forecasting with a probability of 50 %/50 % (randomness) and 80–90 %/100 % (psychological tests) are not without social significance (Ryan, & Ployhart, 2000).

To illustrate, let us recall the procedures of psychological examinations of aircraft operators. The starting point is the development of the so-called pilot's psychological profile, including nonspecific requirements in the scope of optimal operational risk management, psychophysical fitness, and situational awareness, which depend among others on such pilot's resources as health condition, cognitive fitness, personality and operational skills, etc. (Zawadzki, 1931). Attention is paid solely to the procedure and scope of the recruitment tests for air force candidates. Standard psychological examination of military pilots covers the following areas: (a) psychological interview; (b) assessment of intellectual performance (visual memory, attention processes, distribution of visual attention in the central and peripheral field of vision, distribution of attention resources over several tasks, assessment of speed and distance of objects, assessment of susceptibility of perception processes to fatigue); (c) assessment of psychomotor performance (time of simple response and choice, coordination of arms and legs, etc.); and (d) personality assessment (emotional maturity and exclusion of psychopathological traits, social and leadership skills, temperament structure) (Maciejczyk, 2001). Selected tests are used for this purpose, which are exhibited with the use of specialized testing equipment included in the so-called Vienna Testing System (Schuhfried, 1994; Bauer et al., 2002; Byrdoff, 1993).

Moreover, in the case of tests of specific operator efficiency (e.g., in the case of training a pilot for a specific type of aircraft), i.e., selection tests, various aviation simulators are used, which are helpful in diagnosing the structure of operative efficiency in conditions similar to real-world conditions. For example, in the study of spatial imagination in dynamic test conditions, so-called special aviation instruments (e.g., gyroscope, Rhönrad, etc.) (Krawczyk et al., 2017) and a specialist "GYRO-LAB" type simulators (Lewkowicz, R. 2016) are used. Moreover, the results of selection tests using aviation simulators and psychological tests in most cases result in a high correlation coefficient (p=0.01 and p=0.05) with the ratings of flight instructors in real flight (Maciejczyk. 2001).

However, no static tests of spatial imagination training can replace airplanes-laboratories in which both the spatial imagination and the balance organ found in the middle ear can be trained under the control of flight instructors while performing spatial aerial figures in real flight.

Concluding the part of the book devoted to the operator–machine–computer relationship, it is necessary to realize that the modern operator of technical devices such as military and passenger aircrafts, submarines, spacecrafts, etc., due to computer control and visualization moves precisely in three-dimensional real space, due to virtual computer models. An example is the International Space Station, to which spacecraft operators dock with great precision.

As it results from the previous deliberations in the psychology of the operator of technical devices, the discussed problems of the role of the operator in terms of the system are identified as an important factor of reliability of the occupational safety system, in which the so-called intelligent machines are increasingly involved, which are equipped with "artificial intelligence", improving the quality of life of modern people while still maintaining their special role in the O–M system (Terelak, 2011).

6. Computer as a virtual operator serving the optimization of the quality of life

In the postmodern period in which we live in, the computer, as an "intelligent" tool of the man, has ceased to be just a machine and has often become an independent virtual operator of micro- and nano-devices improving the quality of life. Therefore, it is worth giving some examples illustrating this thesis. Undoubtedly, the basis of the thesis formulated in this way should be connected with the so-called artificial intelligence, the advantages and limitations of which in comparison to the natural intelligence of a man are illustrated by Fig. 20.

As shown in Fig. 20, the fundamental difference between artificial intelligence and natural intelligence of a man is based on a separate structure of reasoning: algorithmic (intelligent machine) vs. heuristic (man). Thus an intelligent machine "behaves" according to the "imprinted" closed procedures, while a human being is guided by open principles, according to which they can generate an infinitive number of procedures. The advantage of the intelligent machine is visible at the sensory level and involuntary memory of humans, who, especially in a situation of limited time, make too many mistakes. This advantage, due to a miniaturized electromechanical system at the "input" to the O–M system, simulates people's natural senses in the form of microsensors, providing people with much more information about the external environment of the system than a person would be able to do, especially when there is a shortage of time or overload on the operating memory. Processing this information requires miniaturized technologies based on the Micro-Electro-Mechanical System (MEMS) and the Nano-Electro-Mechanical System (NEMS), which allow for the construction of technical devices in unimaginably small dimensions (Lyshevsky, 2002; Wannok, 2002). MEMS integrates microscale systems in itself, such as motion, electromagnetic, radiation energy, optical energy, etc. It performs four basic tasks: (1) conversion of physical stimuli to electrical, mechanical, and optical signal form and vice versa; (2) activation of devices, "sensing" surrounding stimuli and other functions; (3) taking control (intelligence, decision making, gradual learning, adaptation, self-organization, etc.); and (4) diagnosis of subsequent states of a device, signaling current and future hazards. MEMS and NEMS are based on Artificial Neural Networks (ANN) composed of simple structural elements (micro- and nano-structures), but capable of independently modifying the properties of the connections between themselves. The task of an artificial neuron, constructed to mimic a biological structure, is to sum up the incoming

Artificial intelligence	Natural intelligence
Automatic processing: - not fully controlled, - runs smoothly, - is holistic, - is subject to gradual influence of the practice, - its modification is difficult, - high level of performance, - introspective low availability.	Controlled processing: - its requires effort, - is partial, - it can be interrupted or initiated at any time, - does not give in to the influence of the practice, - but it is easier to make an ad hoc correction, - the level of performance is relatively low, - the introspective availability of the processed material is high

Fig. 20: Differences between artificial and natural intelligence in operator information processing (own elaborations)

signals, adequate to the modality of the sensor and, after exceeding the threshold value, to send a collective signal to another neuron in a simple or complex system of connections.

ANN has its architecture (detector, location, and relations with other neurons), the process of searching for relevant stimuli in the environment (sensory modality, sensitivity of sensors), as well as a specific method of learning to detect the difference between input data and a given result.

Although artificial neural networks are based on human biological models, although they are a very useful tool, not only in detecting various signals that exceed the natural reception capacity of humans, but above all in optimizing the decision-making processes, especially in situations with a limited amount of time, they differ significantly from natural neural networks in the human brain (cf. Fig. 20). The main difference between Artificial Neural Networks and Natural Neural Networks is that the former react only in situations similar to the "output" ones (as taught and set by the constructor). This means that their "intelligence" does not depend on learning, as in the case of humans, but on advances in micro- and nano-technology.

Some of these technologies, applied to the construction of modern cars and aircraft, illustrate the dynamic progress of technical civilization at the turn of the 20th and 21st century. Here are some examples of parallel application of high technology in airplanes vs. cars: (a) automatic control (autopilot vs. *Intelligent Cruise Control – ICC; Flight Management System – FMS* vs. *Automatic Lane Keeping – ALK*); (b) navigation (*Moving map displays* vs. *Route guidance systems; Flight Management System – FMS* vs. *Routefinders; GPS – Global Positioning System* vs.

GPS – Global Positioning System); (c) improving visibility (*Night Vision Goggles – NVGs vs. Vision enhancement systems; Forward Looking Infrared – FLIR*); and (d) avoiding collision (*Collision avoidance systems vs. Traffic alert and Collision Avoidance System – TCAS; Ground Proximity Warning System – GPWS*).

Therefore, there are many systems which facilitate the handling of aircraft, water and road and increase work safety. It is worth noting that the availability of such systems on the market is growing (e.g. GPS in road and air traffic) (Casner, 2005). At the end of the 20th century, in the *Armstrong Aerospace Medical Research Laboratory* in Wright-Patterson Air Force Base, Ohio, the works on the cockpit devoid of any visual pilot-navigation and pilot motion control devices were completed. Instead of the devices, direct brain activity (electrophysiological waves) coupled via an interface with the aircraft control computer shall be used (Green, Self, Ellifritt, 1995). At the same time, *Servicio Iriscom* has developed a computer control system designed for paralyzed people, using eye motion detection as input data. Looking at any part of the computer screen triggers a cursor reaction, which moves immediately. The "enter key" or "mouse click" are replaced with blinks. If the keyboard is displayed on the screen, a person can type a text, navigate the Internet, etc., using eye motions and blinking[41]. The base of this system is the analysis of eye motions using infrared light reflected from the cornea. This idea was used in the construction of a viewfinder placed on a helmet of the "look-aim-shoot" type (e.g., Israeli DASH – *Display and Sight Helmet*), which allows the pilot's sight to point the missile at the target using an electromagnetic sensor located in the pilot's helmet (the so-called third eye) coupled with the computer system of aircraft navigation, a self-guidance system, and a HUD – *Head-Up Display*.

Although the operational needs of the army (*Combat asymmetric scenario*) and the needs of civil institutions in the fight against terrorism (*Terrorist scenario*) are a major impetus for the development of micro- and nano-technology, modern technology has been adopted quite quickly in everyday public life in terms of improving health (*Health scenario*) and the quality of life (*Well-being scenario*) (Daniel, et al., 2005). The first two groups of needs were mainly related

---

41 So a person with a functioning brain is able to communicate with the environment, e.g., by typing on the screen, even if they are completely paralyzed, as in the case of the famous physicist Stephen Hawking. Despite the proof that Hawking gave us with his life questions: Is our brain able to directly communicate with and control the machine? These are still the subjects of a lot of research in the field of neuroscience.

to the construction of various types of sensors for early and precise detection of sources of physical, chemical, biological hazards, etc. The last two needs were to increase the effectiveness of disease detection and pharmacological therapy (e.g., insulin nanopumps, etc.) as well as to make everyday life easier by giving our unreliable senses the ability to detect various warning signals against different threats (e.g., failure of technical devices, etc.).

6.1 Virtual senses supporting the operator's situational awareness

At the heart of the MEMS and NEMS design is the dynamic development of microprocessors and *sensors requirements*. The development of the former contributed to the construction of multifunction "intelligent" computers and the latter to the optimization of cognitive processes at the sensual level. Modern sensory systems already include: *electro-optical sensors, acoustic sensors, radar sensors, biological sensors, chemical sensors, radiation sensors, mines and explosives sensors, and health sensors* (monitoring, e.g., metabolic processes, temperature, energy expenditure, dehydration, state of functioning of individual organs, etc.). Some of them exceed the detection capabilities of human senses, such as electro-optical (used in night vision goggles) or radiation (e.g., ionizing radiation detectors), etc., in this way broadening the possibilities of human activity in the world and contributing to the creation of artificial environment (*artificial satellites*), due to which man can explore previously inaccessible underwater, terrestrial, and cosmic regions (e.g., artificial habitats in the form of bathyspheres and bathyscaphes, submarines, planes, spacecrafts, space stations, etc.). It is therefore worthwhile to characterize individual types of sensors from the point of view of their practical applications.

6.1.1 Virtual electro-optical sensors

This type of sensors are used in the construction of night vision devices. As we know, unlike night animals (e.g., owls, snakes, etc.), human vision becomes compromised at dusk or at night. Therefore, various generations of night vision devices which improve night vision quality are being developed, for operators of military equipment in particular. For example, the use of night vision significantly improves the visibility for a pilot in flying in the darkness (about 2,000 times) as compared to visibility without night vision enhancements (Prost et al., 2005). Although the use of night vision goggles significantly improves night visibility, it also has negative consequences compared to daylight visibility, i.e., the

Virtual senses supporting the operator 149

transition from binocular to monovision vision, which worsens the precise assessment of distance, size of objects, and their shadows (Hartong et al., 2004).

Similar devices, miniaturized to the size of goggles, called eye-tracking, record saccadic eye movements. They generally consist of goggles connected with two systems: *photo-optical* (generating a light directed at the cornea) and *recording* eye movements (miniature video camera) the indicator of which is a marker of the light reflected from the cornea fixed in a specific region of the field of vision. Mixing both these marker images and fragments of the field of view, the exact position of the eye at the time of measurement can be observed. It is a small, portable device. They can be mounted on the head of a driver, jet pilot, or conveyor belt operator in a factory and record the entire process of receiving visual information under natural working conditions *online*.

6.1.2 Acoustic and radar sensors

Acoustic sensors are used in echolocation both underwater (e.g., submarines) and on the ground. For example, in the military, soldiers have these type of sensors mounted on their own helmets to detect position – *snipers and mortals sensors* (Naz et al., 2007). The location of the detectors at different places on helmets and their sensitivity set at a frequency of about 20 kHz, will locate the source of acoustic waves within a radius of 10 to 200 meters (*bullet's shock wave*) (Scanlon, 2008). In the case of drivers of this type, detectors mounted on the outermost parts of the vehicle (e.g., in bumpers) make it easier to assess the distance from an obstacle while parking.

Infrared targeting and tracking systems are worth mentioning. For example, the IRST – *Infrared Search and Track* – system used in military aviation supplements radar detection, because it is effective and very accurate in detecting targets emitting heat in the form of infrared radiation (*IR signature*) of different wavelengths (depending on the temperature of the object emits heat caused by the friction of air against the surface of a moving object) with a small radar cross-section (e.g., self-steering missiles) and is resistant to active electronic interference

The system of acoustic detectors integrates and analyzes sources located on different platforms (vehicles, robots, soldiers, planes), providing, e.g., on a modern battlefield, information optimizing staff decisions, such as the deployment of snipers, artillery fire control, cooperation with tanks and helicopters. This system, coupled with another system, called GPS or *Global Positioning System*, also allows you to locate an aircraft and even the position of a single soldier on the battlefield (if equipped with an acoustic helmet). I am thinking here of the E-3, the American early warning and control plane equipped with an airground network-centric communication system commonly known as AWACS – *Airborne Warning and Control System*. In addition, modern armies are equipped *helmet-mounted bullet-tracking radar* systems (with built-in radio wave sensors), which is coupled with AWAKS or GPS (Bernstein et al., 1998).

6.1.3 Biological, chemical, and radiation sensors

Biological sensors are used primarily to minimize biological hazards, which are either related with terrorist attacks or environmental disasters. Detecting and locating sources of biological hazards is difficult for two reasons: (1) microorganisms are omnipresent in the world around us and (2) the environment is extremely complex and dynamic. Therefore, the construction of biological sensors requires interaction between such fields of science as meteorology, physics, chemistry, and biology. Since bacteria, viruses, and other microorganisms form highly complex biological communities, *Biological Agents – BA* – must not only detect and distinguish between individual microorganisms, but also their pathogens on many planes of reality (water, air, food, and other biologically active liquids). *BA* consists of three interconnected systems: detection, transducer, and analyzer. The detection system is based on bioreceptors capturing molecules from the environment through the use of various technologies: biochemical (e.g., enzyme monitoring), immunological (e.g., monitoring of hormones, drugs, antibodies, viruses, toxins, etc., using an immunochromatograph), nucleic acids (e.g., DNA testing), tissues and cells, chemical, and physical. In general, global analysis includes primarily biosensors based on: biochemistry, enzymes, immunology, and nucleic acids. The system processing biosensory components is mainly based on electrochemical (e.g., amperometry, potentiometry, impedance, conductometry), optical (e.g., reflectance, spectroscopy, fluorescence, chemiluminescence), thermal, and calorimetric technologies. Analytical system and interpreting biological data flowing into it cooperates with a failure detector in the system. The entire biological platform requires cooperation with a high-end computer equipped with comprehensive software (Rodriguez-Mozas, Marco, Lopez de Alda, 2004).

Radiation sensors, commonly referred to as "dirty bomb" detectors, are based on technologies of fast location of radiation sources and analysis of various radioactive materials both in nature and technology (e.g., production of isotopes, nuclear power plants, radiological laboratories in hospitals, etc.), as well as in the fight against terrorism (illegal trade in fissile materials). As we know, radiation cannot be detected with the human senses, which is why portable detection devices have been used for many years, such as Geiger counters, Geiger-Müller counters, dosimeters, etc. Radiation sensors register gamma waves emitted by radioactive substances. There are two indicators: (1) quantitative – *Radiation Absorbed Dose – Rad;* (2) qualitative – *Roentgen Equivalent Man – Rem.* The doses absorbed by the body accumulate and, depending on the time of exposure, may be more or less harmful and lead to radiation sickness (Tubiana, 2003).

6.2 Medical applications of a sensory platform

The advances in micromedicine and nanomedicine are due to the needs of the American army, associated with the need to monitor the health and exercise capacity of soldiers on the battlefield. Clinical medicine benefits from this first one, while sports medicine benefits from the latter (Greorge, 2004). This is connected with the implementation of two programs: *The Warfighter Physiological Status Monitoring (WPSM)* and *Future Force Warrior (FFW – US Soldiers of the Future Program).* The WPSM *system* monitors, among other things: (1) vital parameters, e.g., heart rate, breathing, blood pressure, body temperature, etc.; and (2) exercise and fatigue parameters, e.g., metabolism (three data categories: basal, thermal effect, exercise effect), sugar levels, dehydration, and mental states. An example is the clinical version of a computer-controlled medical device for the assessment of basic vital parameters – heart activity and environmental parameters. The device makes a digital recording of electrocardiographic parameters (the portable part with sensors mounted on the chest) in combination with the physical flight parameters: temperature, altitude, and acceleration values (the part installed in the aircraft), on the basis of which the workload level and the so-called psychophysiological cost of the operator in real working conditions are determined (part of the equipment analyzing the data sent *online* to a computer is located in a medical center). The literature mentions a number of such miniaturized recorders, even up to the size of nanospheric chips (Różanowski, Dziuda, Skibniewski, 2006).

The most extensive development to date has been achieved in sugar level monitoring systems, which in the USA and Europe has seen the construction of six sensor systems and portable devices: (1) Continuous Glucose Monitoring System Gold (CGMS Gold) (sensor-based electrochemical enzyme analysis performed every 5 minutes for 72 hours); (2) GlucoWatch G2 Biographer – GlucoWatch – is based on autosensors of skin generating low levels of electricity and iontophoresis reaction, measuring every 10 minutes for 13 hours; (3) Guardian Telemetered Glucose Monitoring System and Guardian RT – Guardian – a system similar to CGMS, but with the exception that if the blood sugar level is too high or too low, it generates an alarm; (4) GlucoDay-S – GDS – semi-invasive blood glucose monitoring system, based on minidialysis technology, has a micropump in a portable kit (2 cm tube placed in the skin and a microsensor); (5) Short Term Sensor – STS – is a continuation of the previous system and consists in placing a mini applicator and monitoring sensor in the skin for 72 hours; and (6) The FreeStyle Navigator – based on electrochemical sensory elements technology. Some of these systems transmit data online to the monitoring and analysis unit, therefore their further development goes in two directions: qualitative and quantitative. As far as the qualitative direction is concerned, it consists in simultaneous monitoring of various vital health parameters, such as the portable, polyparametric *AMON Integrated System* or the *Field-Hospital-on-a-Chip concept* (Kane, 2008). The quantitative direction of development concerns the further miniaturization of diagnostic and therapeutic apparatus, such as a silicon chip placed on a fingertip (2 x 2 mm) developed by the Oak Ridge National Laboratory (USA) or a thermometer (*The Pill-Thermometer*), placed in a silicone capsule (the size of a pill) with crystalline thermosensors inside that when swallowed, transmit by radio waves the body thermal values every 15 seconds for 12–48 urs http://www.nasa.gov/vision/earth/technologies/thermometer pill.html - 2007).

We mentioned an example of the development of modern microtechnology (MEMS), which has a practical application in medicine, and sparked a competition, in a sense, between biotechnology companies and led to the development of NEMS nanotechnology, as exemplified by the construction of an extremely small insulin nanopump (60 x 40 mm/30 g) manufactured by *Debiotech SA and STMicroelectronics* (Piveteau, 2007).

The development of nanomedicine has been a fact for several years now, especially in the field of remote surgery. The development of nanomedicine promises to eliminate the surgeon from surgical operations entirely, as the work on the construction of miniaturized surgical instruments (nanotools) miniaturized to the size of a pill is now complete. A pre-programmed nanosurgical procedure can be started by the patient by pressing "enter". The first generation of medical minirobots, penetrating the human body through the digestive, circulatory, and respiratory systems, has been announced by the *Institute of Robotics and Intelligent Systems* in Zürich.

The areas of application of micro- and nano-technology in medicine concern many human organs, such as: (1) *Brain* – (a) Sensory helmet that picks up brain waves and allows a person with limb paralysis to use their hand; (b) A spinal cord prosthesis implanted into the motor cortex, recording the activity of individual neurons, sends signals bypassing the spinal cord to the wrist muscles, moving the paralyzed part of the hand; (d) The "lie detection helmet" which scans the brain

Applications of the sensor platform 153

using functional magnetic resonance to determine truthfulness more accurately than traditional polygraphic tests; (d) The electrodes implanted directly into the nerve cells of the brain just below the cranium using electrocorticography (ECoG) allow paralyzed people to use a computer on their own. (Data from implants or electrodes are sent via fiber optic cable to a decoder where the nerve impulses are converted into digital form and then the data from the decoder is sent via cable to the computer); (2) *Eye* – Platinum electrodes, microcameras and ultrasonic distance detectors and a retina built of photodiodes to stimulate a layer of large ganglion cells in the visual cortex, generating a black and white image; (3) *Ear* – implanted in the ear (cochlea) is a microphone with a processor placed on the ear transmitting processed sounds directly to the brain stem; (4) Heart – "Artificial heart" made of plastic and titanium implanted or extracorporeal pump or "replaceable parts" such as heart valves or pacemakers; (5) Arm – A bionic part moved by impulses from the nervous system; (6) *Lung* – Consists of a microbaloon that increases the oxidation capacity of the blood 300 times/min; and (7). *Diaphragm* – Electrodes implanted near the ends of diaphragm nerves and outside the body connected to a stimulator allow people with spinal cord injury to breathe.

From the previous considerations, it results, among other things, that although artificial sensors are the basis for the development of micro- and nano-mechanics and new directions of biotechnology, it is only through the interface with a PC that they create multidimensional systems allowing for the integration of the obtained data, contributing to the development of new fields of technical civilization (e.g., robotics, microbiotechnology, micro- and nano-medicine, molecular physics, etc.) (Kopczynski, Młynczak, Nyga, et al., 2009).

6.3 Applications of the sensor platform – Interface in the development of high technologies optimizing the quality of work and life

As has been discussed so far, the use of *high technology* in civil life, although comes with some delay compared to military purposes, contributes to the improvement of the quality of life of modern people. However, the further development of MEMS and NEMS depends on a proper platform for the integration of sensory (multisensory) data. Theoretically, the sensor platform – interface can be imagined as a multilevel processing of the obtained sensory data with online feedback in very short periods of time. This requires a very wide access to identification of data through modern means of communication (means of communication, much faster and have a higher capacity than classical means, such as: intel system, sonar, radar, GPS), with global and cosmic range. A multistep analysis of the data collected should include five steps: (1) Stage Zero – is indicative of the areas of reality from which the information samples originate; (2) Stage One – assessment of the essence of the signal in terms of its strength, physical characteristics, direction of action, identification, and recognition; (3) Stage Two – concerns situational conditions and the importance of reaching the information system; (4) Stage Three – is related to the initial assessment of the anticipated operation of the system; and (5) Stage Four – is intended to optimize data integration by making appropriate adjustments due to the interaction of this block with both the computer and by obtaining the necessary additional data from external links to the integration system. All previously mentioned stages of the system, from 1 to 4, are connected through an interface with a computer, equipped with a block of collecting and maintaining data in the data library memory and with a block that integrates the incoming data *online*. The effect of cooperation of all platform's blocks is the corrective action of the fourth stage. Corrected and optimized data are entered into the reprocessing, either at the stage level or directly into the Central Computer. Thus, processed and grouped data are obtained at the "output" of the platform, which in turn can be sent to various servers managing metadata even in geospace (e.g., *Open Geospatial Consortium*). The architecture of such a server includes sensory modules, from which data is transmitted via GPS to the server of high data integration for identification (*IDCP – Identification Data Combining Process*) and use in action[42].

Concluding this part of the book, it should be emphasized that at the turn of the 20th and 21st century, in the field of operator activity, man made a huge leap forward. Continuous improvement of work tools, the apogee of which is the construction of an "intelligent machine" in the form of a new generation of computers, allows people not only to improve the quality of everyday life, but above all to conduct scientific research on the surrounding reality in three dimensions: macrocosmic, geophysical, and microcosmic. Exploration of terrestrial regions (military, telecommunications, meteorological, and geological satellites; space stations; astronautics; expeditions to the Moon and Mars)[43] and outer

---

42 IDCP is currently a standard NATO countries (STANAG 4162) and is used for commercial aviation project to optimize communication between aircraft, approaching ground stations and other flying objects (aircraft - GPS – interface - server).

43 For example, the Atlas V rocket launched towards the Sun at the beginning of February 2010 brought the Solar Dynamics Observatory into space. In addition to observing solar activity, the satellite is equipped with innovative on-board instrument

Applications of the sensor platform 155

space (space probes)[44] in addition to cognitive curiosity, which is the basic need for intellectual development of the human species, also brings specific terrestrial benefits (e.g., satellite communications, optimization of weather forecasts, development of modern rocket aviation, development of microprocessor technology, improvement of laser technology, etc.). Exploring the microcosm provides similar benefits, which, unlike the macrocosm, does not even have a very rich science fiction literature, because this world exceeds our imagination. Currently, due to the development of micro- and nano-technology it is not only possible to penetrate our body at the cellular level (e.g., brain mapping, nanomedicine, nanopharmacology, etc.), but through the progress of molecular physics and nanophysics, it is also possible to penetrate the inside of the atom, photographing protons moving at a huge speed in phantoseconds[45], or electrons circulating on orbits, etc. Remaining constantly coupled with a computer interface, provided an intelligent tool that qualitatively changes the face of the modern world, which not only replaced our unreliable cognition with artificial sensors, but also optimized many of our decisions, especially those that required analyzing excessive amounts of data (in relation to our working memory) in a short time.

---

constructions, which on-line will send data to the Earth at 150 million bits per second, creating images ten times better than HD television.

44 Launched in 1977, the Voyager 1 and Voyager 2 spacecraft have reached the edge of the Solar System and are moving to the interstellar space, still sending data to Earth, thanks to which we learned the composition of the atmosphere of large planets, mainly composed of hydrogen and helium. Earlier fired Pioneer 10 and Pioneer 11 probes (sent to space in 1970) after exploring Jupiter and Saturn, are still moving into open space, having on their decks a plate with an engraved engraving depicting the location of the Earth, technical documentation of the probe, and figures of a man and a woman addressed to possibly encountered representatives of another civilization.

45 You can imagine a Fantosecond as a 32 millionth part of one second.

Final thoughts

Summarizing the entirety of the book on the situation of people at work, characterized from the perspective of the postmodern times, it should be emphasized that under the influence of dynamically developing technical work tools, the image of the operator has radically changed in terms of their professional, health, and psychological qualifications, and thus the subject of interest in the psychology of work. Machines manufactured in the technocentric period as instruments of operator's action initially generated some difficulties in mastering them, being a source of stress for many operators. However, in the anthropocentric phase, they not only facilitate production, but gradually take over many mental functions (e.g., sensors, computer-aided decision-making processes, etc.), originally reserved exclusively for humans in the O–M system. Computer-controlled "intelligent machines" not only increase work safety (e.g., automation of technological lines in hazardous working conditions) and comfort (e.g. air-conditioning, artificial environment in passenger aircraft, submarines, and spacecraft), but also fulfill the ancient dreams of the mythological Icarus related to flying through the air and space exploration (development of astronomy, cosmonautics) (Wallace, 2004). M. Tarafdar, C.L. Cooper, and J-F. Stich, (2019) point this out, calling the O–M relationship a "techno-stress", defining it as a mechanism explaining the emotional states that operators experience when they are supported by information systems. Information systems designed to make work easier for operators are called *Techno-Eustress*. If these systems do not support the operator's work or even make it more difficult, they call the issue *Techno-Distress*. Both types of *Techno-stress* are the subject of scientific research within the psychology of work, trying to answer questions about how and why the use of Information Systems (IS) causes operators to experience different emotions (negative vs. positive) and what determines these differences. This requires highly specialized interdisciplinary research. The results of this research are passed on to ergonomists and machine designers in order to eliminate "techno-distress".

Examples of such cooperation between psychology, medicine, and technical sciences have been observed since the second half of the 20th century in the field of exploration of both the macrocosm and the microcosm (the human body) on the scale of miniaturization of technical tools to micro- and nano-technology, unimaginable so far in the development of technical civilization, which has found application in, e.g., micro- and nano-medicine. The latter area in particular allows for a real penetration of the human body for purposes such: (a) surgical operations that are more efficient, faster and safer for non-invasive surgery (cf. e.g. the Australian microrobot named *Proteus* or the microrobot American *HeartLander*, which replaces traditional and endoscopic surgery); s (b) pharmacotherapy (dosage of the drug, e.g., in diabetes through a "cyberpill" placed in chips, circulating in the body); (c) diagnostics (e.g., the use of the diagnostic chips inside the body); (d) direct stimulation of specific parts of the brain (e.g., not only in treating Parkinson's as it has been thus far, but also in treating mental illnesses, depression, or obsessive-compulsive disorders); and (e) rehabilitation in the case of limb prostheses connected directly to motor centers of the brain (e.g., artificial arms or legs allowing to experience the feeling of warmth, cold, touch, or even pressure). The opinion of a MIT biologist Eric Drexler, formulated in the book entitled *Engines of Creation*, that the near future of nanomedicine will see miniature "machines" repairing DNA errors associated with the aging process, will not be an exaggeration, although it may still be closer to fantasy today. Certainly, these and other fantasies, concerning, e.g., banks and stem cell cloning, despite the fact that they raise moral doubts (but not in all European countries) (Kopania, 2002), will be possible in the 21st century, which will certainly no longer be called the postmodern era, but colloquially speaking the "super-modern" era. Dreams of immortality are as old as humanity itself, which proves that people in their cognitive curiosity do not admit to the limits of mental cognition (tackled by epistemology and cognitive science, among others), especially that in the postmodern times, through the interface with a computer, as well as the improvement of computers themselves, people gained an "intelligent" partner for self-creation and "changing the Earth". So far, this self-creation is promoted in the concept of cyberhumans, which will be enhanced with various types of micro- and nano-chips: (1) microdevices enhancing senses (e.g., long distance, twilight, and night vision); (2) microsensors for comprehensive organ diagnosis; (3) nanorobots in blood vessels for analysis of blood composition and blood pressure; (4) real-time automatic translation of all languages (microtranslator); (5) biocybernetic implants to activate mental processes and memory; (6) stimulators in brain sensory centers to simulate real sensations, such as pleasure, satiety, various smells; (7) a brain-to-Internet connection that facilitates online access to vast computer database resources and the ability to communicate with anyone at any given time; and (8) Verichip (2.1 mm x 12 mm) implanted under the skin (under local anesthesia) will provide a form of patient identification for physicians (e.g., blood group, disease history, DNA), cash cards in ATMs, identification cards, etc. The Cyberhuman can also be equipped with chips implanted into different parts of "intelligent clothing": (1) digital glasses equipped with GPS, displaying the image of the surrounding area, distance to the target, and the position of objects in relation to each other; (2) display screen sewn into the sleeve of an "intelligent jacket", transmitting the latest news, electronic letters, videos, etc.; and (3) devices mounted in the trouser legs, enhancing muscle strength, etc.

Perhaps the envisioned image of a cyberhuman is too fantastic, but the descriptions of underwater adventures of Captain Nemo and his "Nautilus", created by Verne's imagination, were also perceived as purely fantastical, and yet during my life they already became reality. Perhaps the 21st century is rightly described by Mirosław Stańczyk in an article entitled "*Man of the twenty-first century*", where in neobiblical language he wrote: "Paralyzed people shall start walking, the blind shall see, diabetics shall forget about insulin injections. People with heart attacks shall receive new hearts and patients with damaged liver shall not wait for a transplant" (Wprost, Nr ½, 2010 published on January 3, 2010, p. 75). We already know today that a new era of medicine is upon us, connected with tissue engineering and organ breeding.

Anticipating further development of new technologies in the 21st century in relation to the intellectual development of human beings, we enter the field of futurology, in which, in the opinion of the outstanding inventor and scientific dreamer – Ray Kurzweil, the psychological-biological model of man, which has been promoted in psychology – must be transformed into a modern psychobiological-cybernetic model in the near future. This new model of the man of the future, known as the idea of *Singularity*, is already the basis of education at the *Singularity University* co-founded by Peter Diamandis, where leading theorists and practitioners from fields such as artificial intelligence, nanotechnology, biotechnology, advanced computer techniques, etc., teach alongside philosophers, psychologists, and neurophysiologists.

List of figures

Fig. 1: The basic model of the operator–machine (O–M) (own elaborations)) 22
Fig. 2: Accuracy of reading the scale depending on the shape of the instrument's dial (own elaborations) 30
Fig. 3: The intelligent CSE (own elaborations based on Hollnagel (2001)) 55
Fig. 4: Virtual Panoramic Display (by Instrument Flying Handbook (2001): courtesy of the Archive of the Military Institute of Aviation Medicine) 57
Fig. 5: Models for performing two simultaneous tasks (own calculations based on: Szczechura, Malawski (1999)) 66
Fig. 6: Basic elements of the executive attention of the operator of technical devices (own calculations) 69
Fig. 7: Directional orientation systems (source: own elaborations) 70
Fig. 8: Model of Posner's cognitive cycle at the eye-tracking level (own elab. base on Russo, 1978) (Legend: S - perception of the Stimulus, A – detachment Attention from the previous stimulus, D - Decision about eye movement, C - Counting eye movement parameters, M - eye Movement; Normal cycle - slow: S-A=D-C-M; Normal cycle – fast: S-D-C-M; E$xpress cycle: S-C-M). 76
Fig. 9: Frequency of eye fixation and the proportions of the time of eye fixation (in sec.) on individual instruments pilot-navigating during the start (own elaborations) 80
Fig. 10: The order and the part of the pilot's fixation of the pilot performing the "engineless autorotation" emergency maneuver on the MI-2 helicopter (own research) 83
Fig. 11: Mechanism of sensory correction of targeted movements (own study based on: Nazarow, 1969) (A - the beginning of the movement, B - the starting point of the stabilizing correction, C - the final point of the movement) 90
Fig. 12: The procedure of analysis of the left-hand coordination structure from the eyecentric perspective (own calculations based on the Howard model - 1982) 93
Fig. 13: The 1970s Bryans SMA-3 static simulator for examining visual-motor coordination (own archive) 101

List of figures

Fig. 14: The structure of the trajectory of eye movements reflecting individual differences in cognitive styles: conservative vs. anticipatory (own research) 102

Fig. 15: The 1960s Dynamic Link simulator used to test the pilot's effectiveness (own archive, courtesy of the Archive of the Military Institute of Aviation Medicine) 104

Fig. 16: Four stages of development of avionics in the 20th century, *significantly changing the aircraft operator's work* (own study) 117

Fig. 17: Three levels of error of operators of technical devices (Own elaboration) 132

Fig. 18: The model of operator errors: stochastic (A) and systematic (b) (Own elaboration) 136

Fig. 19: Model of the structure of the psychological diagnosis in the situation of recruitment (select in) and secondary selection (select out) (own study) 141

Fig. 20: Differences between artificial and natural intelligence in operator information processing (own elaborations) 146

List of table

Tab. 1: Health and psychological risks of exposure to certain neurotoxins 51

Literature

Abrams, T.S., Martin, C.D., Orr, C.E., Hinson, T.A. (1991). Cockpit automation technology CSERIAC-CAT Jul 89 – Dec 90: Final Report No. AL-TR-1991-0078. Wright-Patterson AFB, OH: Armstrong Laboratory. (DTIC No. A273124).

Ackers, P. (2006). The history of occupational psychology: A view from industrial relations. Journal of Occupational and Organizational Psychology, 79, 213–216.

Akerstedt, T., Lindbeck, G. (2007). Night shift work. In: G. Fink (ed.): Encyclopedia of Stress. Second Edition (Vol. 2, pp. 917–920). New York: Academic Press.

Albery, W.B. (2004). Acceleration in other axes affects +Gz tolerance: Dynamic centrifuge simulation of agile flight. Aviation, Space, and Environmental Medicine, 75(1), 1–6.

Anderson, N., Ones, D.S., Sinangil, H.K., Viswesvaran, C. (2001). Handbook of Industrial, Work, and Organizational Psychology, Vol. 1–2. London: SAGE Pub. Ltd.

Anderson, S.W., Rizzo, M., Shi, Q., Uc, E.Y., Dawson, J.D. (2005). Cognitive abilities related to driving performance in a simulator and crashing on the road. In: Paper presented at the Third International Driving Symposium on Human Factors in Driver Assessment, Training and Vehicle Design, Maine, USA.

Andrzejuk, A. (2006). Uczucia i sprawność: Związek uczuć i sprawności w „Summa Theologiae" św. Tomasz z Akwinu. (Emotions and condition: The connection between emotions and condition in St. Thomas Aquinas's "Summa Theologiae".) (p. 50–51) Warszawa: NAVO.

Anokhin, P.K. (1970). Teoria funkcjonalnoj sistemy. In: W.A. Trapeznikov (ed.): Obszczyje woprosy fizjołogiczeskich mechanizmow (p. 6–43). Moscow: Nauka.

Averty, P., Collet, C., Dittmar, A., Athenes, S., Vernet-Maury, E. (2004). Mental workload in air traffic control: An index constructed from field test. Aviation, Space and Environmental Medicine, 75(4), 333–341.

Avison, W.R. (2007). Environmental factors. In: G. Fink (ed.): Encyclopedia of Stress. Second Edition (Vol. I, pp. 934–940). New York: Academic Press.

Bańka, A. (1998). Poglądy Wojciecha Jastrzębowskiego na pracę i naukę o pracy (Labour and science of labour as viewed by Wojciech Jastrzebowski). Ergonomia, 21, 25–37.

Bańka, A., Chirkowska-Smolak, T. (1994). Charakterystyka zawodów i ofert pracy (Characteristics of Occupations and Job Offers). Poznań: Print-B.

Barton, R.R. (1988). G-induced loss of consciousness: Definition, history, current status. Aviation, Space, and Environmental Medicine, 59, 2–5.

Bauer, H., Guttmann, G., Leodolter, M., Leodolter, U. (2002). Sensorimotor Coordination – SMK. Wien: Modling, Dr G. Schuhfried Ges. M.B.H.

Bernstein, U., Lefevre, R., Mann, J., Avent, R., Deo, N. (1998). Sniper bullet detection by milimeter-wave radar. Proceedings of SPIE, Vol. 3577, pp. 231–242.

Biegeleisen, B. (1931). Berufswünsche der Jugend in Krakau (Polen). W: P. F. Lazarsfeld (red.): Jugend und Beruf (s. 130–139). Jena: Verlag G. Fischer.

Biegeleisen-Żelazowski, B. (1964). Zarys psychologii pracy (Outline of the Psychology of Work). Warszawa: PWN.

Biela, A., Kamiński, L., Manek, A., Pietraszkiewicz, H., Sienkiewicz, Z., Szumilewicz, J. (1992). Kwestionariusz Lubelski Stanowiska Pracy (Lublin Job Posts Questionnaire). Lublin: Publishing house of the Lublin Catholic University.

Biernacki, M., Bicka-Capała, M., Tarnowski, A. (2007). Teoretyczne i metodologiczne problemy badania obciążenia na przykładzie metody subiektywnego obciążenia pracą. (Theoretical and methodological problems of researching strain, based on the example of the subjective workload method.) The Polish Journal of Aviation Medicine, Bioengineering and Psychology, 13(4), 465–479.

Björklund, C.M., Alfredson, J., Dekker, S.W. (2006). Mode monitoring and call-outs: An eye tracking study of two-crew automated flight deck operations. The International Journal of Aviation Psychology, 16(3), 263–275.

Black, F.O. (2001), What can posturography tell us about vestibular function? Annale New York Academy Sciences, 942, 446-464.

Blake, M.J. (1967). Relationship between circadian rhythm of body temperature and introversion/extraversion. Nature, 215, 896–897.

Blatteis, C.M. (2007). Thermal stress. In: G. Fink (ed.): Encyclopedia of Stress. Second Edition (Vol. 3, pp. 723–726). New York: Academic Press.

Bobniewa, (1969). Problem niezawodności człowieka. (The problem of human reliability.) In: Z. Kapuścińska, J. Okóń (eds.): Psychologia inżynieryjna w ZSRR i USA (Engineering Psychology in the USSR and the USA) (pp. 35–43). Warszawa: Wydawnictwo Książka i Wiedza.

Bonnes, M., Secchiaroli, G. (1995). Environmental Psychology. London: SAGE Pub. Ltd.

Borowsky, B., Oron-Gilad, T. (2016). The effects of automation failure and secondary task on drivers' ability to mitigate hazards in highly or semi-automated vehicles. Advances in Transportation Studies an International Journal, Special Issue, 1(59), 60–70.

Bourrez, A. (1999). Developing Turn-key ATM systems in Europe. In: Proceedings of the ODT Workshop: Lessons Learnt by ATC Providers and Industry Suppliers. Brussels (Belgium), 11–12 February, 1999, pp. 115–121.

Bracken, D.W. (2007). Review of recruiting, interviewing, selecting & orienting new employees. Personnel Psychology, 60(4), 1074–1077.

Brannick, M.T., Levine, E.L., Morgeson, F.P. (2007). Job and Work Analysis: Methods, Research, and Applications for Human Resource Management. Second Edition. London: SAGE Pub. Ltd.

Breaugh, J.A., Starke, M. (2000). Research on employee recruitment: So many studies, so many remaining questions. Journal of Management, 26, 405–434.

Bruce, V., Green, P.R., Georgeson, M.A. (1996). Visual Perception: Physiology, Psychology, Ecology. Third Edition. Hove: Erlbaum.

Buckley, T.C., Blanchard, E.B. (2007). Motor vehicle accidents, stress effects of. W: G. Fink (ed.): Encyclopedia of Stress. Second edition (Vol. 2, pp. 764–767). New York: Academic Press.

Buskist, W.F., Davis, S.E. (eds.) (2007). 21st Century Psychology: A Reference Handbook. London: SAGE Publ. Inc.

Byrdoff, P. (1993). Pilot candidate selection in the Royal Danish Air Force and EURO-NATO Air Crew Selection Working Group Battery and Research Program. The International Journal of Aviation Psychology, 1, 379–383.

Cable, D.M., Edwards, J.R. (2004). Complementary and supplementary fit: A theoretical and empirical integration. Journal of Applied Psychology, 89(5), 822–834.

Cacciabue, P.C. (1999). Modelling and simulation of human behaviour in process control: Needs, perspectives and applications. In: D. Harris (ed.): Engineering Psychology and Cognitive Ergonomics: Job Design, Product Design and Human-Computer Interaction (Vol. 4, pp. 3–20). Aldershot: Ashgate.

Caplan, R.D. (1987). Person-environment fit theory and organizations: Commensurate dimensions, time perspectives, and mechanisms. Journal of Vocational Behavior, 31, 248–267.

Carless, S.A. (2003). A longitudinal study of applicant reactions to selection procedures and job and organizational characteristics. International Journal of Selection and Assessment, 11, 345–351.

Carless, S.A. (2005). Person–job fit versus person–organization fit as predictors of organizational attraction and job acceptance intentions: A longitudinal study. Journal of Occupational and Organizational Psychology, 78, 411–429.

Carreta, T. (1993). Pilot candidate selection in the US Air Force. The International Journal of Aviation Psychology, 1, 384-388.

Carretta, T.R., Ree, M.J. (2003). Pilot selection methods. In: M.S. Tsang, M.A. Vidulich (eds.): Principles and Practice of Aviation Medicine (pp. 357-396). London: Lawrence Erlbaum Associates, Publishers.

Casner, S.M. (2005). The effect of GPS and moving map displays on navigational awareness while flying under VFR. International Journal of Applied Aviation Studies, 5, 153-165.

Chapanis, A. (1961). Men, machines and models. American Psychologists, 16, 113-131.

Chapanis, A. (1963). Engineering psychology. Annual Review of Psychology, 14, 285-350.

Chapanis A. (1969). Psychologia inżynieryjna, w: Z. Kapuścińska, J. Okóń (red.), Psychologia inżynieryjna w ZSSR i USA (Engineering psychology. In: Z. Kapuścińska, J. Okóń (ed.), Engineering psychology in the USSR and USA (pp. 254-307). Warszawa: Książka i Wiedza.

Cheveigne, A. de. (2001). The auditory system as a "separation machine." In: A.J.M. Houtsma, A. Kohlrausch, V.F. Prijs, R. Schoonhoven (eds.): Physiological and Psychophysical Bases of Auditory Function (pp. 393-400). Maastricht: Shaker.

Chmiel, N. (ed.) (2000). An Introduction to Work and Organizational Psychology. A European Perspective. Oxford: Blackwell Publishers.

Christensen, J.M., Talbot, J.M. (1986). A review of the psychological aspects of space flight. Aviation, Space and Environmental Medicine, 57(3), 203-212.

Clamann, M., Kaber, D.B. (2004). Applicability of usability evaluation techniques to aviation system. The International Journal of Aviation Psychology, 14(1), 395-420.

Cockell, C.S., Bush, T., Bryce, C., Direito, S., Fox-Powell, M., Harrison, J.P., Lammer, H., Landenmark, H., Martin-Torres, J., Nicholson, N., Noack, L., O'Malley-James, J., Payler, S.J., Rushby, A., Samuels, T., Schwendner, P., Wadsworth, J. and Zorzano, M.P. (2016). Habitability: A review. Astrobiology, 16 (1), 1-29.

Cohen, S., Spacapan, S. (1984). The social psychology of noise. In: D.M. Jones, A.J. Chapman (eds.): Noise and Society (pp. 221-245). New York: Wiley.

Colquhoun, W.P. (ed.) (1971): Biological Rhythms and Human Performance (s. 39-107). London: Academic Press.

Convertino, V.A. (2007). Space, health risk of. In: G. Fink (ed.): Encyclopedia of Stress. Second Edition (Vol. 3, pp. 548-554). New York: Academic Press.

Cook, M.J., Wilson, K., Proctor, L.J. (2001). Disruptive effects between multi-modal tasks. In: D. Harris (ed.): Engineering Psychology and Cognitive Ergonomics: Industrial Ergonomics, HCI, and Applied Cognitive Psychology (Vol. 6, pp. 263-269). Aldershot: Ashgate.

Crawford, D. (1998). Environments and adaptations: Then and now. In: C.C. Crawford, D.L. Krebs (eds.): Handbook of Evolutionary Psychology: Ideas, Issues, and Applications (pp. 275-302). Mahwah: Erlbaum.

Czekaj, J., Ćwiklicki, M. (2009). Infonomics - a New Science? E-mentor, 29(2), 4-7.

Dahleberg, A. (2001). Air rage: The under estimated safety risk. Aldershot: Ashgate.

voren, M., McCauley, D. (2007). Work-related stress. Psychiatric Bulletin, 31(8), 316-317.

Dekker, S., Rigner, J. (1999). Training for the automated task: Investigating axpertise for modern flight decks. In: D. Harris (ed.): Engineering Psychology and Cognitive Ergonomics: Transportation Systems, Medical Ergonomics and Training (Vol. 3, pp. 249-257). Aldershot: Ashgate.

DeVera, J-P., Boettger, U., de la Torre Noetzel, R., Spohn, T. (2012). Supporting Mars exploration: BIOMEX in Low Earth Orbit and further astrobiological studies on the Moon using Raman and PanCam technology. Planetary and Space Science, 74, 103-110.

De Voge, J.M., Bass, E.J. (2007). Job-specyfic or general training: A quantitative assessment. The International Journal of Aviation Psychology, 17(4), 333-351.

De Voogt, A., Van Doorn, R.A. (2007). The paradox of helicopter emergency training. The International Journal of Aviation Psychology, 17(3), 265-274.

Di Majo, V., Rebessi, S., Pazzaglia, S., Saran, A., Covelli, V. (2003). Carcinogenesis after low doses of ionizing radiation. Radiation Research, 159, 102-108.

Domański, H., Sawinski, Z., Słomczyński, K.M. (2007). Sociological Tools Measuring Occupations: New Classification and Scales. Warszawa: Publishing House of the Institute of Philosophy and Sociology of the Polish Academy of Sciences.

Dudek, B., Makowska, Z. (1986). Skala do pomiaru psychologicznego obciążenia pracą (Psychological work-load measurement scales). Przegląd Psychologiczny, 29(2), 541-554.

Duffey, R.B., Saull, J.W. (2008). Managing Risk: The Human Element. Chichester (UK): John Wiley and Sons, Ltd., Pub.

Dunnette, M.D., Hough, L.M. (eds.) (1990-1992). Handbook of Industrial and Organizational Psychology (Second edition) (Vol. 1-3). Palo Alto: Consulting Psychologists Press, Inc.

Edwards, E. (1979). Human error. Proceedings of Symposium „Dutch Air-Line Pilots Association: Safety and efficiency – The next 50 years, The Hague, 1–7 Septembr, 1979.

Edwards, E. (1981). Safety via discipline. Proceedings of Symposium "Safety management", Trondheim, 8–10 January, 1981, Norwegian Society of Charterad Engineers.

Edwards, J. R. (1991). Person-job fit: A conceptual integration, literature review, and methodological critique. In: C.L. Cooper, I.T. Robertson (eds.): International Review of Industrial and Organizational Psychology (Vol. 6, pp. 283–357). New York: Wiley.

Ek, Å., Akselsson, R. (2007). Aviation on the ground: Safety culture in a ground handling company. The International Journal of Aviation Psychology, 17(1), 59–76.

Ely, J.H., Thomson, R.M., Orlansky, J. (1956). Design of Controls: The Joint Services Human Egineering Guide to Equipment Design. Chapter VI. WADC Technical Report 56–172. Wright-Patterson AFB, Ohio: Wright Air Development Center.

Enander, A. (1984). Performance and sensory aspects of work in cold environments: A review. Ergonomics, 27(4), 365–378.

Fąfrowicz, M., Marek, T. (1999). Werońska koncepcja źródeł stresu. (Verona concept of sources of stress). In: J.F. Terelak (ed.): Źródła stresu: Teoria i badania. (Sources of Stress. Theory and Research.) (p. 13–22). Warszawa: Published by ATK.

Filding, N., Folding, J. (1986). Linking Data. Newbury Park: Sage Publications.

Fischer, B. (1986). The role of attention in the preparation of visual guided eye movements in monkey and man. Psychological Research, 48, 251–257.

Fischer B., Weber H. (1993). Express saccades and visual attention. Behavioral and Brain Sciences, 16(3), 553–610.

Fowler, B., Prilic, H., Brabant, M. (1994). Acute hypoxia fails to influence two aspects of short-term memory: Implications for the source of cognitive deficits. Aviation, Space, and Environmental Medicine, 5, 641–645.

Fowler, B., Taylor, M., Porlier, G. (1987). The effect of hypoxia on reaction time and movement time components of a perceptual-motor task. Ergonomics, 30(10), 1475–1485.

Franus, E. (1977). Model niezawodności człowieka i jego znaczenie dla ergonomii. (Human reliability model and its importance for ergonomics). Przegląd Psychologiczny, 20(1), 35–47.

Franus, E. (1978). Myślenie techniczne (Technical Thinking). Wrocław: Ossolineum.

Friedman, G. (1960). Maszyna i człowiek: Problem człowieka w cywilizacji maszynowej. (Industrial Society: The Emergence of Human Problems of Automation) (orig. Fr. Machine et humanisme: Problèmes humains du machinisme industriel). Warszawa: Wydawnictwo Książka i Wiedza.

Gaździński, S.P. (2017). Hemodynamic parameters and brain oxygenation in military pilots as a function of acceleration's duration at 4G and at 6G: A preliminary study. The Polish Journal of Aviation Medicine, Bioengineering and Psychology, 23(2), 5–10.

Goemaere, S., Brenning, K., Beyers, W., Vermeulen, A.C.J., Binsted, K., Vansteenkiste, M. (2019). Do astronauts benefit from autonomy? Investigating perceived autonomy supportive communication by Mission Support, crew motivation and collaboration during HI-SEAS 1. Acta Astronautica, 157, 9–16.

Green, R.G., Self, H.S., Ellifritt, T.S. (eds.) (1995). 50 years of Human Engineering: History and Cumulative Bibliography of the Fitts Human Engineering Division. Springfield: National Technical Information Service.

Green, R.G., Muir, H., James, M., Gradwell, D. Green, R.L. (1996). Human Factors for Aircrew. Aldershot: Ashgate.

Gregore, A.K. (2004). Comparison of activity monitors to estimate energy cost of treadmill exercise. Medicine and Science in Sports and Exercise, 36(7), 1244–1251.

Guardiera, S., Schneider, S., Noppe, A., Strüder, K. (2008). Motor performance and motor learning in sustained + 3 Gc acceleration. Aviation, Space, and Environmental Medicine, 79(9), 852–858.

Guilford, J.P. (1988). Some changes in the Structure – of – Intellect Model. Educational and Psychological Measurement, 48, 1–4.

Haddon, W. (1980). Options for the prevention of motor vehicle crash injure. Israel Journal of Medicine Science, 16, 45–68.

Haakonson, N.H. (1980). Investigation of life change as a contributing factor in aircraft accidents: A prospectus. Aviation, Space and Environmental Medicine, 51(9), 981–988.

Hancock, P.A., Desmond, P.A. (eds.) (2001). Stress, Workload and Fatigue. Mahwah: Lawrence Erlbaum Associates, Publishers.

Hannum, K.M., (2008). Review of evaluating human resources programs: A 6-phase approach for optimizing performance. Personnel Psychology, 61(2), 459–462.

Harris, D. (ed.) (1999). Engineering Psychology and Cognitive Ergonomics: Transportation Systems, Medical Ergonomics and Training. Vol. 3–4. Aldershot: Ashgate.

Harris, D. (ed.) (2001). Engineering Psychology and Cognitive Ergonomics: Industrial Ergonomics, HCI, and Applied Cognitive Psychology. Vol. 6. Aldershot: Ashgate.

Hart, S.G., Staveland, L.E. (1988). Development of NASA-TLX (Task Load Index): Results of empirical and theoretical research. In: P.A. Hancock, N. Meshkati (eds.): Human Mental Workload (pp. 239–250). North-Holland: Elsevier Science Publishers.

Hartong, D.T, Jorritsma, F.F, Neve, J.J., Melis-Dankers, B.J.M., Kooijman, A.C. (2004). Improved mobility and independence of night-blind people using night-vision goggles. Investigative Ophthalmology & Visual Science, 45, 1725–1731.

Haslam, S.A. (2007). Psychology in Organizations. Second Edition. London: SAGE Pub. Ltd.

Hawkins, F.H. (1993). Human Factors in Flight. Second Edition. Aldershot: Ashgate.

Hebb, D.O. (1965) Drives and the C.N.S. In: H. Fowler (ed.): Curiosity and Exploratory Behavior (pp. 176–190). New York: McMillan Co.

Heinrich, H.W. (1959). Industrial Accident Prevention: A Scientific Approach. New York, Toronto, London: McGraw-Hill Book Company, Inc.

Hermaszewski, M. (2013). Psychological aspects of space flights. The Polish Journal of Aviation Medicine and Psychology, 9(3): 5–8.

Heyer, M. (1997). Introduction to infonomics. The Institute for Infonomics Research Silicon Valley World Internet Center, http://www.heyertech.com/document/Infonomics.pdf.

Hobbs, A., Kanki, B.G. (2008). Patterns of error in Confidentional Maintenance Incident Reports. The International Journal of Aviation Psychology, 18(1), 5–16.

Holland, J.L. (1997). Making Vocational Choices: A Theory of Vocational Personalities and Work Environments. Third Edition.. Odessa: Psychological Assessment Resources.

Hollnagel, E. (2001). The user in control: From HMI to JCS. In: D. Harris (ed.): Engineering Psychology and Cognitive Ergonomics: Industrial Ergonomics, HCI, and Applied Cognitive Psychology (Vol. 6, pp. 3–11). Aldershot: Ashgate.

Horne, J.A., Östberg, O. (1976). A self-assessment questionnaire to determine morningness-eveningness in human circadian rhythms. International Journal of Chronobiology, 4, 97–110.

Houston, C.S., Sutton, J.R., Cymerman, A., Reeves, J. (1987). Operation Everest II: Man at extreme altitude. Journal of Applied Physiology, 63(2), 877–882.

Howard, J.P. (1982). Human Visual Orientation. Chichester: Wiley.

Howard, J., Cunningham, D., Rechnitzer, P. (1978). Stress and Survival on the Job: Rusting Out, Burning Out, Bowing Out. Toronto: Macmillan Book.

Hull, J.G., Draghici, A.M., Sargent, J.D. (2012). A longitudinal study of risk-glorifying video games and reckless driving. Psychology of Popular Media Culture, 1(4), 244–253.

Instrument Flying Handbook (2001). Oklahoma City: U.S. Department of Transportation Federal Aviation Administration, Flight Standard Service.

Isaac, A.R., Ruitenberg, B. (1999). Air Traffic Control: Human Performance Factors. Aldershot: Ashgate.

Ivanov, V., Ilyin, L., Gorski. A., Tukov, A., Naumenko, R. (2004). Radiation and epidemiological analysis for solid cancer incidence among nuclear workers who participated in recovery operations following the accident at the Chernobyl NPP. Journal of Radiation Research (Tokyo), 45, 41–44.

Jacob, P., Jeannerod, M. (2003). Ways of Seeing: The Scope and Limits of Visual Cognition. Oxford: Oxford University Press.

Jacob, R.J.K., Karn, K.S. (2002). Commentary on Section 4. Eye tracking in human-computer interaction and usability research: Ready to deliver the promises. Authors' preprint, (jacob@cs.tufts.edu).

Jamroz, K. (2008). Review of road safety theories and models. Journal of KONBiN, 1(4), 89–108.

Jasiński, T.L. (2005). Znaczenie ukierunkowanego treningu fizycznego w zwiększaniu tolerancji organizmu pilota wojskowego na przyspieszenia +Gz. (The importance of targeted physical training in increasing the tolerance of the air force pilot's to +Gz accelerations.) Studia Monograficzne AWF (University of Physical Education – Monographic Studies) No. 32. Kraków: Wyd. University of Physical Education in Kraków.

Jederberg, W.W., Still, K. (2002). The utilization of risk assessment in tactical command decisions. The Science of the Total Environment, 288, 119–129.

Jensen, R.S. (ed.) (1989). Aviation Psychology. Aldershot: Gower Technical.

Jethon, Z. (1976). Działalność operatorowa – nowa postać pracy człowieka. (Operator Activity – A New Form of Human Work). Warszawa: PWN.

Johnson, B.L., Anger, W.K. (1982). Behavioral toxicology. W: W.N. Rom (eds.): Environmental and Occupational Medicine. London: Little Brown.

Jones, L.V. (2007). Some lasting consequences of US psychology programs in World Wars I and II. Multivariate Behavioral Research, 42(3), 593–608.

Kaber, D.B., Perry, C.M., Segall, N., Sheik-Nainar, M.A. (2007). Workload state classification with autumation during simulated Air Traffic Control. The International Journal of Aviation Psychology, 17(4), 371–390.

Kane, D. (2008). Field-Hospital-on-a-Chip Project Awarded to NanoEnginieer from San Diego. San Diego News Centre. Available from: http://ucsdnews.ucsd.edu/newsrel/science/10-08FieldHospitalOnAChip.asp

Karp, M.R., Condit, D., Nullmeyer, R. (1999). Survey of Cockpit/Crew Resource Management for F-16 Pilots (AFHRL-HE-AZ -TR-1999-XXXX). Mesa: Air Force Research Laboratory, Warfighter Training Research Division.

Kinsbourne, M. (1981). Single channel theory. W: D. Holding (red.): Human Skills (pp. 65–85). New York: John Wiley & Sons.

Klein, K.E., Wehmann, H.M., Hunt, B.I. (1972). Desynchronization of body temperature and performance circadian rhythm as a result of outgoing and homegoing transmeridian fights. Aerospace Medicine, 43 (2), 119–132.

Klein, K.E., Wehmann, H.M., Hunt, B.I. (1980).Significance of circadian rhythms in aerospace operations. North Atlantic Treaty Organization, Advisory Group for Aerospace Research and Development (AGARDograph No. 247). DFVLR-Institut fur Flugmedizin 5300 Bonn-Bad-Godesberg FRG.

Kocian, D.F. (1991). Visually Coupled Systems (VCS): The Virtual Panoramic Display (VPD) "system". In: K. Krishen (ed.): Fifth Annual Workshop on Space Operations, Applications, and Research (NASA CP-3127, Vol. 2, pp. 548–561). Washington, DC: National Aeronautics and Space Administration.

Konorski, J. (1969). Integracyjna działalność mózgu (Integration of the Brain). Warszawa: PWN.

Koonce, J.M. (2002). Human Factors in the Training of Pilots. London: Taylor and Francis.

Kopania, J. (2002). Etyczny wymiar cielesności. Kraków: Aureus.

Kopczynski, K., Mlynczak, J., Nyga, P., et al. (2009). Advanced Helmet and Devices for individual protection (AHEAD): Sensors requirements – WP200 – System definition. Internal.

Koradecka, D. (red.) (1997). Bezpieczeństwo pracy i ergonomia (Work Safety and Ergonomics). Warszawa: CIOP.

Kotarbiński, T. (1955). Traktat o dobrej robocie (Treaty of Good Work). Wrocław: Ossolineum.

Kotarbiński, T. (1960). Sprawność i błąd (Efficiency and error). Warszawa: PZWS.

Kowalczuk, K., Puchalska, L., Palonek, H., Sobotnicki, A., Janewicz, M., Wyleżoł, M., Gaździński, S.P. (2017). Hemodynamic parameters and brain oxygenation in military pilots as a function of acceleration's duration at 4G and at 6G: A preliminary study. The Polish Journal of Aviation Medicine, Bioengineering and Psychology, 23(2), 5–10.

Kowalewska, S. (1965). Definicje i klasyfikacje zawodów. (Definitions and classifications of occupations). In: A. Sarapata (ed.): Socjologia zawodów (Sociology of Occupations; pp. 52–62). Warszawa: Wydawnictwo Książka i Wiedza.

Kowalski, W. (1982). Adaptacja człowieka do pobytu w Antarktyce. (Adaptation of a Man to Stay in Antarctica). Warszawa: Wyd. WIML.

Kowalski, W. (2002). Kabiny ciśnieniowe. W: W. Kowalski (red.): Medycyna lotnicza: Wybrane zagadnienia (Pressure chamber). In: W. Kowalski (ed.): Aviation medicine: Selected issues (pp. 30–153). Poznań: Wyd. DWL i OP.

Krawczyk, P., Kobos, Z., Marcinkowski, P., Jędrys, R., Wochyński, Z. (2017). Physical fitness and the level of learning by pilot cadets in the initial period of studying in the Polish Air Force Academy (PAFA) in Dęblin. The Polish Journal of Aviation Medicine, Bioengineering and Psychology, 23(1), 15–21. DOI: 10.13174/pjambp.19.06.2018.02.

Kristof, A.L. (1996). Person-organization fit: An integrative review of its conceptualizations, measurement, and implications. Personnel Psychology, 49, 1–49.

Kristof–Brown, A.L., Zimmerman, R.D., Johnson, E.C. (2005). Consequenses of individuals' fit at work: A meta – analysis of person – job fit, person – organizaton, person – group fit ang person – supervisors fit. Personnel Psychology, 58(2), 281–342.

Kroemer, K.H., Snook, S.H., Meadows, S.K., Deutsch, S. (eds.) (1988). Ergonomic Models of Anthropometry Human Biomechanics and Operator-Equipment Interfaces. Washington, DC: National Academy Press.

Kubiczkowa, J. (2000). Test statokinezjometryczny w ocenie stanu równowagi (Statokineziometric test in the assessment of the state of equilibrium), (habilitation dissertation). Warszawa: Wyd. CKP-WAM.

Kuliński, J., Prost, M., Stasiak, K., Jezierski, M., Latek-Najder, M. (2005). Sprwność widzenia w goglach noktowizyjnych PNL-3 na podstawie testów kontrstu (Visibility in PNL-3 goggles on the basis of contrast tests). The Polish Journal of Aviation Medicine, Bioengineering and Psychology, 12(2), 137–143.

Kulwicki, P.V., McDaniel, J.W., Guadanga, L.M. (1987). Advanced development of a cockpit automation design support system. AGARD Conference Proceedings 417: The Design, Development, and Testing of Complex Avionics Systems (pp. 19-1–19-15). Neuilly sur Seine, France: NATO Advisory Group for Aerospace Research and Development.

Kuznetsova, P.G., Gushchin, V.I., Vinokhodova, A.G., Chekalina, A.I., Shved, D.M. (2017). Interpersonal interaction under the conditions of high autonomy in interplanetary mission simulation (Mars-500 Experiment). Human Physiology, 43(7), 751–756.

Kwarecki, K., Terelak, J. (1980). Medycyna i psychologia kosmiczna. (Medicine and Space Psychology). Warszawa: Wiedza Powszechna (Omega series).

Kwarecki, K., Święcicki, W., Kłossowski, M., Terelak, J., Zużewicz, K. (1982). Zdolność do pracy fizycznej i umysłowej w warunkach 24–godzinnej bezsenności oraz pracy zmianowej. (Ability to work physically and mentally under conditions of 24 hours sleeplessness and shift work). Medycyna Lotnicza, 74(1), 7–14.

Kwiatkowski, R., Duncan, D.C., Shimmin, S. (2006). Occupational psychology history. Journal of Occupational and Organizational Psychology, 79, 83–201.

Łaszczyńska, J. (2002). Wpływ warunków środowiska termicznego na organizm pilota. (The impact of thermal environment conditions on pilot's body.) In: W. Kowalski (ed.): Medycyna lotnicza: Wybrane zagadnienia (Aviation Medicine: Selected Issues (pp. 152–184). Poznań: Published by DWL i OP.

Lauver, K.J., Kristof-Brown, A. (2001). Distinguishing between employee's perceptions of person-job and person-organization fit. Journal of Vocational Behavior, 59, 454–490.

Lempereur, I., Lauri, M.A. (2006). The psychological effects of constant evaluation on airline pilots: An exploratory study. The International Journal of Aviation Psychology, 16(1), 113–133.

Leuba, C. (1965). Toward some integration of learning theory: The concept of optimal stimulation. In: H. Fowler (ed.): Curiosity and Exploratory Behavior (pp. 169–175). New York: Macmillan Co.

Levine, E.L., Spector, P.E., Menon, P.E., Narayanon, L., Cannon-Bowers, J. (1996). Validity generalization for cognitive, psychomotor, and perceptual test for craft jobs in the utility industry. Human Performance, 9, 1–22.

Lewkowicz, R. (2016). A modelling approach to the human perception of spatial orientation. The Polish Journal of Aviation Medicine, Bioengineering and Psychology, 22(4), 27–42.

Li, W.C., Harris, D. (2008). The evaluation of the effect of a Short Aeronautical Decision-Making Training Program for Military Pilots. The International Journal of Aviation Psychology, 18(2), 135–152.

Lim, D.V., Simpson, J.M., Kearns, E., Kramer, M.F. (2005). Current and developing technologies for monitoring agents of bioterrorism and biowarfare. Clinical Microbiology Reviews, 18(4):583-607. DOI: 10.1128/CMR.18.4.583-607.2005

Locke, E.A. (ed.) (2003). Postmodernism and Management: Pros, Cons, and the Alternative. New York: JAI Press.

Łomow, B.F. (ed.) (1977). Osnowy inżeniernoj psichołogii. Moscow: Wyzszaja Szkoła.

Łomow, B.F., Płatonow, K.K. (red.) (1984). Eksperymentalna psychologia lotnicza (Experimental Psychology in Aviation and Aeronautics). Warszawa: PWN.

Lyshevsky, S.E. (2002). MEMS and NEMS: Systems, Devices, and Structures. Boca Raton: CRC Press.

Maciejczyk, J. (2001). Kwalifikacja psychologiczna kandydatów do szkolenia lotniczego (Psychological selection of candidates for aviation training). The Polish Journal of Aviation Medicine, Bioengineering and Psychology, 7(1), 26–39.

Maciejczyk, J., Biernacki, M. (2005). Przestrzenne przetwarzanie informacji w badaniach pilotów (Spatial information processing in pilot studies). The Polish Journal of Aviation Medicine, Bioengineering and Psychology, 11(4), 401–407.

Maciejczyk, J., Kuzak, W., Skibniewski, F.W. (1996). Ocena zależności między wybranymi testami psychologicznymi i efektywnością ćwiczeń na symulatorze lotu (Assessment of the relationship between selected psychological tests and the effectiveness of training on a flight simulator). The Polish Journal of Aviation Medicine, Bioengineering and Psychology, 2(2), 137–144.

Maciejczyk, J., Terelak, J.F. (2017). Recolection of aviation psychologists from the perspective of the Jubilee of the 90th Anniversary of the Military Institute of Aviation Medicine in Warsaw. The Polish Journal of Aviation Medicine, Bioengineering and Psychology, 23(3–4), 63–73 (DOI: 10.13174/pjambp.).

Martin, W.L. (1992). Developing Virtual Cockpits. AGARD Proceedings of the 63 Avionics Panel Meeting: Advanced Aircraft Interfaces: The Machine Side of the Man-Machine Interface (pp. 8-1–8-8). Neuilly sur Seine: NATO Advisory Group for Aerospace Research and Development.

Mathews, G. (1988). Morningness-eveningness as dimension of personality, state, and, psychophysiological correlates. European Journal of Personality, 2, 277–293.

McCabe, D.P., Roediger, H.L., McDaniel, M.A., Balota, D.A., Hambrick, D.Z. (2010). The relationship between working memory capacity and executive functioning: Evidence for a common executive attention construct. Neuropsychology, 24, 222–243.

McCormick, E.J. (1957). Human Engineering. London: McGraw-Hill Book Company, Inc.

McCormick, E.J. (1964). Antropotechnika: Przystosowanie konstrukcji maszyn i urządzeń do człowieka. Warszawa: WNT.

Meglino, B.M., Ravlin, E.C., DeNisi, A.S. (2000). A meta-analytic examination of realistic job preview effectiveness: A test of three counterintuitive propositions. Human Resource Management Review, 10, 407–434.

Meshkati, N., Hancock, P.A., Rahimi, M. (1995). Techniques in mental workload assessment. W: J.R. Wilson, E.N. Corlett (eds.): Evaluation of Human Work: A Practical Ergonomics Methodology. London: Taylor – Francis.

Migdał, K., Terelak, J. (1986). Uczenie się koordynacji wzrokowo-ruchowej jako predyktor efektywności praktycznego szkolenia lotniczego. (Learning how to achieve visual-motor coordination as a predictor of the effectiveness of practical aviation training). Medycyna Lotnicza, 91(2), 59–66.

Mikuliszyn, R. Żebrowski, M. (2002). Patofizjologia przyspieszeń. (Pathophysiology of accelerations.) In: W. Kowalski (ed.): Medycyna lotnicza. Wybrane zagadnienia (Aviation Medicine: Selected Issues) (pp. 98–137). Poznań: Publishing DWL & OPK.

Missiuro, W., Zawadzki, B. (1928). Psychotechnika w lotnictwie. (Psychotechnique in Aviation.) Warszawa: Inspektorat Lotnictwa (Aviation Inspectorate).

Molińska, M. (2015). Hypoxia exposure and working memory performance: A meta-analysis. The Polish Journal of Aviation Medicine and Psychology, 21(1), 10–19.

Moray, N. (1999). The psychodynamic of human-machine interaction. In: D. Harris (ed.): Engineering Psychology and Cognitive Ergonomics (Vol. 4, pp. 225–235). Aldershot: Ashgate.

Mulder, M., Van der Vaart, H.J.C. (2006). An information-centered analysis of the Tunnel-in-the-Sky Display. Part 3: The interaction of curved trajectories. The International Journal of Aviation Psychology, 16(1), 21–49.

Münsterberg, H. (1914). Grundzüge der Psychotechnik. Leipzig: Barth.

Nagel, D.C. (1988). Human error in aviation operation. In: E.L. Wiener, D.C. Nagel (eds.): Human Factors in Aviation. San Diego: Academic Press.

Navon, D., Miller, J.O. (1987). Role of outcome conflict in dual task interference. Journal of Experimental Psychology: Human Perception and Performance, 13, 438–448.

Naz, P., Hengy, S., Hamery, O., Marty, C. (2007). Acoustic detection and localization for defense and security application. 19th International Congress on Acoustics, Madrid, 2–7 September, 2007.

Nazarow, A. (1969). Reakcje sensomotoryczne i nawyki ruchowe. (Sensomotor reactions and movement habits.) In: Z. Kapuścińska, J. Okoń (eds.) Psychologia inżynieryjna w ZSRR i USA (Engineering Psychology in the USSR and the USA) (pp. 231–246). Warszawa: Książka i Wiedza.

Nebylitsyn, W.D. (1969). Niezawodność pracy operatora w złożonym układzie sterowania. (Reliability of the operator in a complex control system.) In: Z. Kapuścińska, J. Okoń (eds.): Psychologia inżynieryjna w ZSRR i USA (Engineering Psychology in the USSR and the USA) (pp. 44–59). Warszawa: Książka i Wiedza.

Nelson, R.J., Martin II, L.B. (2007). Seasonal changes in stress responses. In: G. Fink (ed.): Encyclopedia of Stress. Second Edition (Vol. 3, pp. 427–431). New York: Academic Press.

Nibbelke, R.J., Emmerson, P., Leggatt, A.P., Hughes, T., Biggin, K., Starr, A. (1999). Human centered design process in the Advenced Flight Deck Technology Project. In: D. Harris (ed.): Engineering Psychology and Cognitive Ergonomics: Transportation Systems, Medical Ergonomics and Training (Vol. 3, pp. 93–100). Aldershot: Ashgate.

Nieznański, M., Obidziński, M. (2019). Verbatim and gist memory and individual differences in inhibition, sustained attention, and working memory capacity. Journal of Cognitive Psychology, 1, 16–33.

Norman, D.A., Draper, S.W. (eds.) (1986). User Centered System Design: New Perspective on Human Computer Interaction. New York: Erlbaum.

Oborne, D.J. (1983). Vibration at work. In: D.J. Oborne, M.M. Grueneberg (eds.): Work and Physical Environment (pp. 151–179). Chichester: Wiley.

Ogilvie, J.R. (2007). Review of effective training: Systems, strategies and practices. Personnel Psychology, 60(4), 1077–1079.

O'Hare, D. (2006). Cognitive functions and performance shaping factors in aviation accidents and incidents. The International Journal of Aviation Psychology, 16(2), 145–156.

Okoń, J. (red.) (1963). Psychologiczne aspekty powstawania błędów w pracy (Psychological aspects of errors at work). Warszawa: Instytut Wydawniczy CRZZ.

Okoń, J., Paluszkiewicz, L. (1963). Psychologia inżynieryjna: Dostosowanie maszyn i urządzeń do człowieka. (Engineering Psychology: Adaptation of Machines and Equipment to Human Beings.) Warszawa: PWN.

Osiński, W. (2000). Anthropomotorics. Poznań: University of Physical Education.

Pacelli, C., Selbmann, L., Zucconi, L., Coleine, C., de Vera, J-P., Rabbow, E., Böttger, U., Dadachova, E., Onofri, S. (2019). Responses of the black fungus *Cryomyces antarcticus* to simulated mars and space conditions on rock analogues. Astrobiology, 19(1), 1–12. DOI: 10.1089/ast.2016.1631.

Paluszkiewicz, L. (1964). Wpływ barwy tarcz i kresek podziałkowych na czytelność wskazań wychyłowych aparatów wskaźnikowych. (The influence of the dials and subdivision lines colors on the clearness of needle indicators readings.) Works of the Central Institute for Labour Protection, 14, 41–60.

Parasuraman, R., Byrne, E.A. (2003). Automation and human performance in aviation. In: M.S. Tsang, M.A. Vidulich (eds.): Principles and Practice of Aviation Medicine (pp. 311–356). London: Lawrence Erlbaum Associates, Publishers.

Park, S. (1987). Human Reliability: Analysis, Prediction, and Prevention of Human Errors. New York: Elsevier.

Pashler, H. (1994). Dual task interference in simple tasks: Data and theory. Psychological Bulletin, 116, 220–244.

Patrick, J. (2003). Training. In: M.S. Tsang, M.A. Vidulich (eds.): Principles and Practice of Aviation Medicine (pp. 397–434). London: Lawrence Erlbaum Associates, Publishers.

Pearson, R.A. (1999). Human factors regulation – from concept to reality. In: D. Harris (ed.): Engineering Psychology and Cognitive Ergonomics: Transportation Systems, Medical Ergonomics and Training (Vol. 3, pp. 27–33). Aldershot: Ashgate.

Piveteau, L.D. (2007). Disposable insulin nanopump from Debiotech and STMicroelectronics marks major breakthrough in diabetes treatment. Debiotech S.A. Available from: http://debiotech.org/news/indeks nw166.html

Pope, A., Bogart, E., Bartolome, D. (1995). Biocybernetic system validates index of operator engagement in automated task. Biological Psychology, 40, 187–195.

Posner, M.I. (1994). Attention: The mechanisms of consciousness. Proceedings of National Academy of Science USA, 91, 7398–7403.

Poulton, E.C. (1976). Continuous noise interferes with work by masking auditory feedback and inner speech. Applied Ergonomics, 7, 79–84.

Poulton, E.C. (1978). Blue collar stressors. In: C. L. Cooper, R. Payne (eds.): Stress at Work (pp. 51–79). New York: John Wiley & Sons.

Prost, M., Jezierski, M., Kuliński, J., Gąsik, M. (2005). Badanie widzenia stereoskopowego w trakcie posługiwani się goglami noktowizyjnymi PNL-3. (The study of stereoscopic vision while using PNL-3 night vision goggles.) The Polish Journal of Aviation Medicine, Bioengineering and Psychology, 11(4), 359–362.

Pylyshyn, Z. (1999). Is vision continuous with cognition? The case for cognitive impenetrability of visual perception. Behavioral and Brain Sciences, 22, 341–365.

Raczek, J. (1993). Koncepcje strukturalizacji i klasyfikacji motoryczności człowieka. (Concepts of structuring and classification of human motility.)

In: W. Osiński (ed.): Motoryczność człowieka – jej struktura, zmienność i uwarunkowania. (Human Motility - Its Structure, Changeability and Conditionings.) Poznań: University of Physical Education.

Radakovic, S.S., Maric, J., Surbatovic, M., Radjen, S., Stefanova, E., Stankovic, N., Filipovic, N. (2007). Effects of acclimation on cognitive performance in soldiers during exertional heat stress. Military Medicine, 172(2), 133–139.

Ratajczak, Z. (1988). Niezawodność człowieka w pracy. (Reliability of People at Work.) Warszawa: PWN.

Ratajczak, Z. (1991). Elementy psychologii pracy. (Elements of the Psychology of Work.) Katowice: University of Silesia.

Reason, J. (1990). Human Error. Cambridge: Cambridge University Press.

Reason, J. (1997). Managing the Risks of Organizational Accidents. Aldershot: Ashgate.

Redfern, P.H., Waterhouse, J.M., Minors, D.S. (1991). Circadian rhythms: Principles and measurement. Pharmacology & Therapeutics, 49(3), 311–327.

Reinhart, R.O. (2008). Basic Flight Physiology. New York: McGraw-Hill Book Company, Inc.

Reykowski, J. (1966). Funkcjonowanie osobowości w warunkach stresu psychologicznego. (The Functioning of Personality under Psychological Stress.) Warszawa: PWN.

Reynolds, S. (1997). Psychological well-being at work: Is prevention better than cure? Journal of Psychosomatic Research, 43(1), 93–102.

Rock, P.B., Harris, M.G. (2006). As a potential control variable for visually guided braking. Human Perception and Performance, 32(2), 251–267.

Rodriguez-Mozas, S., Marco, M.P., Lopez de Alda, M.J. (2004). Biosensors for environmental applications: Future development trends. Pure Application Chemistry, 76(4), 723–752.

Roessingh, J.M. (2005). Transfer of manual flying skills from PC-based simulation to actual flight: Comparison of in-flight measured data and instructor ratings. The International Journal of Aviation Psychology, 15(1), 67–90.

Rogelber, S.G., Reeve, Ch.L. (2007). Encyclopedia of Industrial and Organizational Psychology. London: SAGE Pub. Ltd.

Rollag, K. (2007). Defining the term 'new' in new employee research. Journal of Occupational and Organizational Psychology, 80, 63–75.

Romero, L.M. (2007). Seasonal rhythms. In: G. Fink (ed.): Encyclopedia of Stress. Second Edition (Vol. 3, pp. 432–435). New York: Academic Press.

Roth, J. (2007). Temperature effects. In: G. Fink (ed.): Encyclopedia of Stress. Second Edition (Vol. 3, pp. 717–718). New York: Academic Press.

Różanowski, K., Dziuba, Ł., Skibniewski, W. (2006). System jednoczesnej wieloknałowej rejestracji parametrów elektrokardiogrficznych i środowiskowych VENTUS. (VENTUS - A system of simultaneous multi-channel recording of electrocardiographic and environmental parameters.) The Polish Journal of Aviation Medicine, Bioengineering and Psychology, 12(4), 397–410.

Ruddle, R.A. (2001). Navigation: Am I really lost or virtually there? In: D. Harris (ed.): Engineering Psychology and Cognitive Ergonomics: Industrial Ergonomics, HCI, and Applied Cognitive Psychology (Vol. 6, pp. 135–142). Aldershot: Ashgate.

Russo, J.E. (1978). Adaptation of cognitive processes to the eye movements and the higher psychological functions. W: J.W. Senders, D.F. Fisher, R.A. Monty (eds.): Eye Movements and the Higher Psychological Functions (pp. 89–109). New York: Wiley.

Ryan, A.M., Ployhart, R.E. (2000). Applicants' perceptions of selection procedures and decisions: A critical review and agenda for the future. Journal of Management, 26, 565–606.

Sadowski, B. (2000). Wybrane zagadnienia z fizjologii układu wzrokowego. Seminaria z fizjologii, część I, pod red. E. Szczepańskiej-Sadowskiej i E. Koźniewskiej. Warszawa: Akademia Medyczna w Warszawie, ss 53–79 (Selected issues in the physiology of the visual system. Seminars in physiology, part I, edited by E. Szczepańska-Sadowska and E. Koźniewska. Warsaw: Medical Academy in Warsaw, pp. 53–79).

Salas, E., Cannon-Bowers, J.A. (2000). The anatomy of team training. W: S. Tobias, J.D. Fletcher (eds.): Training and Retraining: A Handbook for Business, Industry, Government, and the Military (pp. 312–335). New York: MacMillan.

Salden, R.J.C., Paas F., Van der Pal, J. (2006). Dynamic task selection in flight management system training. The International Journal of Aviation Psychology, 16(2), 157–174.

Samel, A., Wegmann, H.M. (1989). Circadian rhythm sleep and fatigue in aircrews operating on long-dual routes (pp. 404–422). In: R.S. Jensen (ed.): Aviation Psychology (pp. 342–377). Aldershot: Gower Technical.

Sarapata, A. (ed.) (1965). Socjologia zawodów. (Sociology of Occupations.) Warszawa: Wydawnictwo Książka i Wiedza.

Scanlon, M.V. (2008). Helmet-mounted acoustic array for hostile fire detection and localization in urban environment. Proceedings of SPIE, 6963, 1–13.

Schellekens, J.M., Maijman, T.F. (1999). Evaluation of mental workload by using probe-tasks. In: D. Harris (ed.): Engineering Psychology and Cognitive Ergonomics: Job Design, Product Design and Human-Computer Interaction (Vol. 4, pp. 301–308). Aldershot: Ashgate.

Scherbaum, C.A. (2005). Synthetic validity: Past, present and future. Personnel Psychology, 58, 481–515.

Schmitt, N., Chan, D. (1998). Personnel selection: A theoretical approach. Thousand Oaks, CA: Sage Publications.

Schuhfried, G. (1994). Wiener Test System. Mödling: Schuhfried, GmBh.

Schultz, D., Schultz, S.E. (2008). Psychologia a wyzwania dzisiejszej pracy. (Psychology and Work Today.) Warszawa: Wydawnictwo Naukowe PWN.

Scialfa, C.T., Borkenhagen, D., Lyon, J., Deschênes, M. (2013). A comparison of static and dynamic hazard perception tests. Accident Analysis and Prevention, 51, 268–273.

Sechmenov, I.M. (1986). Odruchy mózgowe. (Reflexes of the Brain.) Warszawa: PWN.

Shannon, C.E., Weaver, W. (1969). The Mathematical Theory of Communication. Chicago: University of Illinois Press.

Sherman, P.J. (2003). Applying Crew Resource Management Theory and methods to the operational environment. In: M.S. Tsang, M.A. Vidulich (Eds.): Principles and practice of aviation medicine (pp. 473-506). London: Lawrence Erlbaum Associates, Publishers.

Simons, H.W., Billig, M. (1994). After Postmodernism: Reconstructing Ideology Critique. London: SAGE Pub. Ltd.

Singleton, W.T. (1979). Safety and risk. W: W.T. Singleton (ed.): The Study of Real Skills: Compliance and Excellence (Vol. 2, pp. 137–154). Lancs: CMTP Press Ltd.

Smallwood, T. (2000). The Arline Training Pilot. Second Edition. Aldershot: Ashgate.

Spencer, P. S. & Schaumburg, H. (1980). Experimental and clinical neurotoxicology. Baltimore: The Williams & Wilkins Company.

Stark, L.L., Lis, S.R. (1981). Scanpaths revisited: Cognitive models direct active looking. W: D.F. Fisher, R.A. Monty, J.W. Senders (eds.): Eye movements: Cognitive and Visual Perception. Hilsdale: Lawrence Erlbaum Associates, Publishers.

Stasiak, K. (2009). Ocena wybranych parametrów sprawności wzrokowej w goglach noktowizyjnych PNL-3 na symulatorze HIPERION. (Evaluation of selected parameters of visual efficiency in PNL-3 night vision goggles on the

HIPERION simulator.) An unpublished doctoral dissertation. Warszawa: Military Institute of Aviation Medicine.

Stephens, Ch.L., Pope, A.L. (2014). Closing the loop between man and machine: Mitigating hazardous states of awareness with adaptive automation. The Polish Journal of Aviation Medicine and Psychology, 20(1), 17–24. DOI: 10.13174/pjamp.20.01.2014.2.

Stokes, G.S., Mumford, M.D., Owens, W.A. (eds.) (1994). Biodata Handbook: Theory, Research, and Use of Biographical Information in Selection and Performance Prediction. Palo Alto: Consulting Psychologists Press, Inc.

Strelau, J. (2006). Temperament jako regulator zachowania z perspektywy półwiecza badań. (Temperament as a Regulator of Behavior: After Fifty Years of Research.) Gdańsk: Gdańskie Wydawnictwo Psychologiczne.

Stubbs, J., Danielsson, M. (2001). Management participation in organisational safety systems. In: D. Harris (ed.): Engineering Psychology and Cognitive Ergonomics: Industrial Ergonomics, HCI, and Applied Cognitive Psychology (Vol. 6, pp. 315–332). Aldershot: Ashgate.

Suedfeld, P., Brcic, J., Johnson, Ph.J., Gushin, V. (2014). Coping strategies during and after space flight: Data from retired cosmonauts. 65th International Astronautical Congress, Canada – 2014. International Aeronautical Federation. IAC-14-A1-22049.

Sundstrom, E. (2008). Review of work in the 21st century: An introduction to industrial and organizational psychology. Personnel Psychology, 61(2), 447–450.

Symon, G., Cassell, C. (2006). Neglected perspectives in work and organizational psychology. Journal of Occupational and Organizational Psychology, 79, 307–314.

Szczechura, J., Malawski, M. (1999). Wykonywanie dwu czynności równocześnie jako źródło stresu. (Performing two activities at the same time as a source of stress.) In: J.F. Terelak (ed.): Źródła stresu: Teoria i badania (Sources of Stress. Theory and Research) (p. 169–183). Warszawa: Wyd. ATK.

Szczechura, J., Terelak, J.F. (1981), Charakterystyka wybranych parametrów fiksacji wzroku w działaniu pilota (Characteristics of selected parameters of eye fixation in the pilot's operation). rzegląd Psychologiczny, 14, 769–775.

Szczechura, J., Terelak, J.F. (1993). Ruchy oczu. (Eye movements.) In: T. Sosnowski, K. Zimmer (ed.): Metody psychofizjologiczne w badaniach psychologicznych (Psychophysiological Methods in Psychological Testing) (p. 157–181). Warszawa: Wydawnictwo Naukowe PWN.

Szczechura, J., Terelak, J., Świątek, H. (1988). Koordynacja wzrokowo-ruchowa jako predyktor wykonania zadania lotniczego. (Visual-motor coordination as a predictor of the performance of an aviation task). Ergonomia, 11(2), 287-292.

Szczechura, J., Terelak, J.F., Kobos, Z., Pińkowski, J. (1998). Oculographic assessment of workload influence on flight performance. International Journal of Aviation Psychology, 8(2), 157-176.

Szmigielski, S. (1974). Chronobiologia. Rytmy biologiczne człowieka. (Chronobiology. Human Biological Rhythms.) Warszawa: PWN.

Szocik, K., Abood, S., Shelhamer, M. (2018). Psychological and biological challenges of the Mars mission viewed through the construct of the evolution of fundamental human needs. Acta Astronautica, 152, 793-799.

Tarnowski, A., Terelak J. (1996). Okoruchowy mechanizm uwagi w sytuacji decyzyjnej. (Oculomotor mechanism of attention in decision-making situation). Czasopismo Psychologiczne, 2(3), 189-194.

Tarnowski, A., Terelak, J.F. (1999). Proste zachowania poznawcze a intencjonalność. (Simple cognitive behaviors and intentionality.) Kognitywistyka i Media w Edukacji, 2(1), 245-266.

Taylor, F.W. (1911). Scientific Management. New York: Harper and Row.

Taylor, M.J. (2007). Hypothermia. In: G. Fink (ed.): Encyclopedia of Stress. Second Edition (Vol. 2, pp. 428-438). New York: Academic Press.

Terelak, J. (1971). Refleksje na temat: Wybór zawodu pilota wojskowego jako czynnik motywacyjny o charakterze neurotyczno-kompensacyjnyn. (Reflections on: Choosing the profession of an air force pilot as a motivational factor of neurotic and compensatory nature.) Medycyna Lotnicza, 33, 83-89.

Terelak, J. (1988). Podstawy psychologii lotniczej. (Basics of Aviation Psychology.) Poznań: DWL.

Terelak, J. (1994). Visual attention and eye-hand-legs coordination in young pilots. In: Proceedings of the 30th International Applied Military Psychology Symposium, Heidelberg (Germany), 25-28 July, 1994.

Terelak, J. (1995). Selective attention and sensomotor performance learning in pilots. In: Proceedings of the 31st International Applied Military Psychology Symposium, Lisboa (Portugal), 15-19 May, 1995, pp. 203-209.

Terelak, J. (2011). Człowiek w sytuacji pracy w okresie ponowoczesności. (Man in a Post-Modern Working Situation.) Warszawa: Wydawnictwo UKSW.

Terelak, J.F. (2002). Zjawisko „blackout" - granica życia i śmierci. (The blackout phenomenon – Life and death limit.) In: W. Bołoz, M. Ryś (ed.): Między życiem a śmiercią: uzależnienia, eutanazja, sytuacje graniczne (Between Life and Death. Addictions, Euthanasia, Boundary Situation) (p. 266-276). Warszawa: Wydawnictwo UKSW.

Terelak, J.F. (2011). Szczególna rola człowieka w układzie Człowiek-Obiekt Techniczny Otoczenie – COTO. (Human particular role in Man-Machine-Environment – MME structure.) Polski Przegląd Medycyny i Psychologii Lotniczej, 17(1), 53–73.

Terelak, J.F. (2015). Psychologia kierowców pojazdów drogowych: Teoria i stan badań. (Psychology of Road Vehicle Drivers: Theory and State of Research.) Warszawa: Wydawnictwo UKSW.

Terelak, J.F. (2016). Człowiek w Kosmosie: Bariery adaptacyjne z perspektywy astronautycznej. (Man in space: Adaptational barriers from the astronautical perspective.) Studia Philosophiae Christianae, 52(3), 111–129.

Terelak, J.F. (2017). Characteristics of the scientific and implementational activities of aviation psychologists and scientific consultancy from the perspective of the 90 years of existence of the Military Institute of Aviation Medicine. The Polish Journal of Aviation Medicine, Bioengineering and Psychology, 23(3–4): 74–87 (DOI: 10.13174/pjambp.).

Terelak, J., Kobos, Z. (1996). The formation of eye-hand co-ordination under the influence of exercise on special aviation gymnastic devices. In: Proceedings of the 32nd International Applied Psychology Symposium, 20–24 May, 1996, Brussels (Belgium).

Terelak, J., Szczechura, J. (1987). Zastosowanie badań okulograficznych w locie do oceny ergonomicznej kabiny śmigłowca MI-2. (Application of oculographic examinations in flight to the ergonomic assessment of the cabin of MI-2 helicopter.) Ergonomia, 10(1), 9–26.

Terelak, J., Szczechura, J. (1998). Ocena stanu psychicznego załogi statku powietrznego przed wypadkiem i po zaistnieniu zagrożenia. (Assessment of the mental condition of aircraft crews before and after an accident.) Polski Przegląd Medycyny Lotniczej, 4(3), 311–320.

Terelak, J.F., Tarnowski, A. (1999). Trudność zadania jako źródło stresu. (The difficulty of a task as a source of stress.) In: J.F. Terelak (ed.): Źródła stresu: Teoria i badania (Sources of Stress. Theory and Research) (p. 142–168). Warszawa: Wyd. ATK.

Terelak, J.F., Truszczyński, O. (2000). Psychologiczne determinanty dezorientacji przestrzennej jako błędu przetwarzania informacji multisensorycznej. (Psychological determinants of spatial disorientation as the error of multisensory information processing.) Zeszyty Naukowe WSSM, 2(2), 61–73.

Terelak, J.F., Tarnowski, A., Dobrowolski, A. (1993). Walidacja Skali Chronotypu Człowieka J.A. Horne'a i O. Ŏstberga jako narzędzia badania typów rannych i wieczornych. (Validation of the J.A. Horne and O. Ŏstberg Human Chronotype Scale as a tool for testing morning and evening types.) Przegląd Psychologiczny, 34(3), 363–378.

Tichon, J. (2007). Training cognitive skills in virtual reality: Measuring performance. Cyber Psychology and Behavior, 10(2), 286-289.

Thierens, H., Vral, A., Barbe, M., Meijlaers, M., Baeyens, A., Ridder, L.D. (2002). Chromosomal radiosensitivity study of temporary nuclear workers and the support of the adaptive response induced by occupational exposure. International Journal of Radiational Biology, 78, 1117-1126.

Thorndike, R.L., Hagen, F. (1959). Ten Thousand Careers. New York: Wiley.

Tichomirow, O.K. (1976). Struktura czynności myślenia człowieka. (Structure of Human Thinking.) Warszawa: PWN.

Tinsley, H.E.A. (2000). The congruence myth: An analysis of the efficacy of the person-environment fit model. Journal of Vocational Behavior, 56, 147-179.

Tomaszewski, T. (1963). Wstęp do psychologii. (Introduction to Psychology.) Warszawa: PWN.

Tomaszewski, T. (1965). O porównywalności zawodów. (On the comparability of occupations.) In: A. Sarapata (ed.): Socjologia zawodów (Sociology of Professions) (pp. 23-51). Warszawa: Wydawnictwo Książka i Wiedza.

Tomaszewski, T. (1968). Człowiek w systemie pracy. (Man in the work system.) W: J. Rosner (red.), Ergonomia. Zagadnienia przystosowania pracy do człowieka (Ergonomics. The Issues of Adapting Work to Man.) (s. 69-122). Warszawa: Wydawnictwo Książka i Wiedza.

Torbjörn, A., Göran, K., Mats, G. (2007). Sleep and sleepiness in relation to stress and dis-placed work. Physiology and Behavior, 92(1-2), 250-255.

Triandis, H.C., Dunnette, M.D., Hough, L.M. (eds.) (1994). Handbook of Industrial and Organizational Psychology (Second Edition), (Vol. 4). Palo Alto: Consulting Psychologists Press, Inc.

Truszczyński, O. (2002). Czynnik ludzki w zdarzeniu lotniczym. (Human factor in an aviation event.) Polski Przegląd Medycyny Lotniczej, 8, 27-33.

Truszczyński, O., Terelak, J. (1996). The influence of altitude hypoxia on psychomotor efficiency of military pilots. In: Proceedings of the 32nd International Applied Military Psychology Symposium, 20-24 May, 1996, Brussels (Belgium).

Truszczyński, O., Nowicki, G., Achimowicz, J. (2012). New telemedical approach to pilot and astronaut psychophysiological status monitoring and impact of augmented cognition on flight safety. The Polish Journal of Aviation Medicine and Psychology, 1(18), 35-43.

Truszczyński, O., Terelak, J.F., Jasiński, T., (2000). Psychophysiological cost of acceleration0 +Gz. In: Proceedings of the 24th Conference of the European Association for Aviation Psychology (EAAP) at Crieff Hydro, Crieff (Scotland), 4-8 September, 2000.

Truszczyński, O., Wojtkowiak, M., Biernacki, M., Kowalczuk, K., Lewkowicz, R. (2012). Effect of high acceleration exposure on visual perception in Polish pilots measured with Critical Fusion Frequency Test (CFFT). The Polish Journal of Aviation Medicine and Psychology, 1(18), 19–27.

Tsang, P.S., Vidulich, M.A. (eds.) (2003). Principles and Practice of Aviation Psychology. New Jersey–London: Lawrence Erlbaum Associates, Publishers.

Tubiana, M. (2003). The carcinogenic effect of low doses: The validity of the linear no-threshold relationship. International Journal of Low Radiation, 1, 1–31.

Turski, B. (2014). Vibration – Hazardous factor in professional environment. The Polish Journal of Aviation Medicine and Psychology, 20(2), 43–48. DOI: 10.13174/pjamp.20.02.2014.5.

UNSCEAR (2000). United Nations Scientific Committee on the Effects of Atomic Radiation. Sources, effects and risks of ionising radiation. Report to the General Assembly, with Annexes. New York: United Nations.

Uszakowa, M. (1969). Czas reakcji i problemy psychologii inżynieryjnej. (Reaction time and problems of engineering psychology.) In: Z. Kapuścińska, J. Okóń (eds.): Psychologia inżynieryjna w ZSRR i USA (Engineering Psychology in the USSR and the USA) (pp. 94–106). Warszawa: Książka i Wiedza.

Værnes, R.J. (2007). Pressure, effects of extreme high and low. In: G. Fink (ed.): Encyclopedia of Stress. Second Edition (Vol. 3, pp.184–194). New York: Academic Press.

Van Paassen, M.M., Mulder, M. (1999). The cognitive ecology of tunnel-in-the-sky displays. In: D. Harris (ed.): Engineering Psychology and Cognitive Ergonomics: Transportation Systems, Medical Ergonomics and Training (Vol. 3, pp. 67–74). Aldershot: Ashgate.

Vegchel, N., De Jonge, J., Landsbergis, P.A. (2005). Occupational stress in (inter) action: the interplay between job demands and job resources. Journal of Organizational Behavior. 26, 535–560.

Vincenzi, D.A., Wise, J.A., Mouloua, M., Hancock, P.A. (2009). Human Factors in Simulation and Training. London: Taylor and Francis, CRC Press.

Von Gierke, H.E., McCloskey, K., Albery, W.B. (1991). Military performance in sustained acceleration and vibration environments. In: R. Gal, A.D. Mangelsdorff (eds.): Handbook of Military Psychology (pp. 335–360). Chichester: John Wiley and Sons.

Walichnowski, W. (2008). Obiektywizacja efektów szkolenia teoretycznego, treningu grawitacyjnego i manewrów przeciw przeciążeniowych pilota w wirówce przeciążeniowej jako ważny czynnik bezpieczeństwa lotów. (Objectivization of the effects of theoretical training, gravitational training and pilot anti-G maneuvers in a high-G centrifuge as an important factor in flight safety.) Polski Przegląd Medycyny Lotniczej, 14(2), 173–182.

Wallace, P. (2004). The Internet in the Workplace: How New Technology is Transforming Work. New York: Cambridge University Press.

Wannok, S. (2002). MEMS (Microelectromechanical System). Available from: http://www.engr.ku.edu/-rhale/ae510/websites f02/mems.pdf

Warren, R. (1993). Total visual scene information for flight. In: E. Trautman (ed.): Vision Topics for Aviation: A New Look at a Traditional Concern. (Report No. IST-DF-93-01, pp. 37–51). University of Central Florida: Institute for Simulation and Training.

Waterhouse, J., Minors, S., Waterhouse, M. (1990).Your Body Clock: How to Live with It, Not against It. Oxford: Oxford University Press.

Wegmann, H.M., Klein, K.E., Conrad, B., Esser, P. (1983). A model for prediction of resynchronisation after time-zone flights. Aviation, Space and Environmental Medicine, 54(6), 524–527.

Wells, M.J., Haas, M.W. (1992). The human factors of helmet-mounted displays and sights. In: M.A. Karim (ed.): Electro-optical Displays (pp. 743–785). New York: Dekker, Marcel.

Westerman, J.W., Cyr, L.A. (2004). An integrative analysis of person-organization fit theories. International Journal of Selection and Assessment, 12(3), 252–261.

Whinnery, J.E., Jones, D.R. (1987). Recurrent +Gz - induced loss of consciousness. Aviation, Space and Environmental Medicine, 58(10), 943–947.

Wickens, C.D. (1980). The structure of attentional resources. W: R. Nickerson (ed.): Attention and Performance (Vol. VIII, 239–257). New York:Erlbaum.

Widerszal-Bazyl, M., Cieślak, R., Derlicka, M., (ed.) (1998). Przewodnik po zawodach (Occupation Guide), Volume I–VII. Warszawa: MPiPS, KUP.

Wiegmann, D.A., Shappell, S.A. (2003). A Human Error Approach to Aviation Accident Analysis. London: Ashgate.

Wilde, G.J.S. (1994). Target Risk: Dealing with the Danger of Death, Disease and Damage in Everyday Decision. Ontario: Queen's University.

Williams, H. (1958). Reliability evaluation of the human component in machine systems. Electrical Manufacturing, 61(4), 78–82.

Wilson, J.R., Corlett, E.N. (eds.) (1995). Evaluation of Human Work: A Practical Ergonomics Methodology. Second Edition. London: Taylor and Francis.

Winter, B.J. (2001). G-tolerance training: Methods, capabilities, and new developments. Ed. Environmental Tectonics Corporation. Aircrew Training Systems, 1–7.

Wojtkowiak, M. (2004). Aktualne metody badań i treningu pilotów w wirówkach przeciążeniowych. (Current methods of testing and training of pilots in high-G centrifuges.) Polski Przegląd Medycyny Lotniczej, 10(4), 373–383.

Wojtkowiak, M. (2013). Selected problems of space medicine. Early physiological researchy at the Military Institute Aviation Medicine. The Polish Journal of Aviation Medicine and Psychology, 19(3), 37–44.

Wojtkowiak, M. (2015). Polish studies on human and animal tolerance to acceleration. The Polish Journal of Aviation Medicine and Psychology, 21(4), 21–33.

Wojtkowiak, M., Szajner, R. (2000). Treningi ktapultowania pilotów w lotnictwie polskim. (Pilot ejection sit training in Polish aviation.) Polski Przegląd Medycyny Lotniczej, 6(2), 128–136.

Wolff, S. (1998). The adaptive response in radiobiology: Evolving insights and implications. Environmental Health Perspectives, 106, 277–283.

Woodworth, R.S., Schlosberg, H. (1963), Psychologia eksperymentalna, Volume 1–2. (Experimental Psychology, Vol. 1–2), Warszawa: Państwowe Wydawnictwo Naukowe.

"Wprost" (2010). January 3, 2010, p. 75.

Żakowska, L., Carsten, O., Jamson, H. (2005). Driver's perception of self explaining road infra-structure and architecture – Simulation study. In. G. Underwood (ed.): Traffic and Transportation Psychology, Theory and Application (pp. 397–405). Amsterdam: Elsevier Ltd.

Zalewska, A. M. (2001). „Arkusz Opisu Pracy" O. Neubergera i M. Allerbeck – adaptacja do warunków polskich. ("Work Description Inventory" by O. Neuberger and M. Allerbeck – Adaptation to Polish conditions.) Studia Psychologiczne, 39(1), 197–217.

Zawadzki, B. (1931). Konstytucja psychofizyczna a zdolność do zawodu lotnika. (The psychophysical constitution and the ability to work as an pilots.) Przegląd Sportowo-Lekarski, 3, 4–7.

Zeidler, W. (2011). Krytyka Tayloryzmu jako kolejny krok w rozwoju metodologii Kurta Lewina. (Critique of Taylorism as the next step in the development of Kurt Lewin's methodology.) In: W. Zeidler, H.E. Lück (eds.): Psychologia europejska w okresie międzywojennym. Sylwetki, osiągnięcia, problemy (European Psychology in the Interwar Period. Silhouettes, Achievements, Problems) (pp. 371–389). Warszawa: VIZJA PRESS & IT.

Zeidler, W. Helmut, L. (2014). Zapomniany dokument: "Księga Pamiątkowa" Pierwszej Ogólnopolskiej Konferencji Psychotechnicznej. Warszawa, styczeń, 1930 (Forgotten document, "Memory Book" of 1st Psychotechnical National Conference, Warsaw, January 1930). Studia Psychologica, 14(2), 77–94.

Zhanga, X., Qua, X., Xuec, H., Taoa, D., Lic, T. (2019). Effects of time of day and taxi route complexity on navigation errors: An experimental study. Accident Analysis and Prevention, 125, 14–19.

Zickar, M.J. (2004). An analysis of industrial-organizational psychology's indifference to labor unions in the United States. Human Relations, 57(2), 145–167.

Zieliński, P., Drozdowski, R., Biernacki, M.P. (2014). Hypoxia and cognitive performance. The Polish Journal of Aviation Medicine and Psychology, 20(4), 5–10.

Zinczenko, W., Majzel, M., Nazarow, A., Cwietkow, A. (1969). Analiza pracy operatora. (Analysis of the operator's work.) In: Z. Kapuścińska, J. Okóń (eds.): Psychologia inżynieryjna w ZSRR i USA (Engineering Psychology in the USSR and the USA) (pp. 17–34). Warszawa: Książka i Wiedza.

Index

A
Abood, S. 48, 185
Abrams, T.S. 55, 165
Achimowicz, J. 130. 187
Ackers, P. 14, 18, 165
Akerstedt, T. 50, 165
Akselsson, R. 135, 170
Albery, W.B. 39, 46, 112, 165, 188
Alfredson, J. 119, 166
Anderson, N. 20, 165
Anderson, S.W. 100, 165
Andrzejuk, A. 21, 165
Anger, W.K. 49, 173
Anokhin, P.K. 128, 165
Athenes, S. 165
Avent, R. 166
Averty, P. 130, 165
Avison, W.R. 35, 165

B
Baeyens, A. 187
Balota, D.A. 177
Bańka, A. 16, 138, 165, 166
Barbe, M. 187
Bartolome, D. 130, 180
Barton, R.R. 47, 56, 166
Bass, E.J. 119, 169
Bauer, H. 143, 166
Bernstein, U. 150, 166
Beyers, W. 171
Bicka-Capała, M. 130, 166
Biegeleisen-Żelazowski, B. 14, 16, 166
Biela, A. 139, 166
Biernacki, M. 46, 108, 130, 166, 177, 188, 191
Biggin, K. 179
Billig, M. 13, 183

Binsted, K. 171
Björklund, C.M. 119, 166
Black, F.O. 71, 166
Blake, M.J. 52, 166
Blanchard, E.B. 50, 167
Blatteis, C.M. 36, 111, 166
Bobniewa, 128, 166
Boettger, U. 48, 169
Bogart, E. 130, 180
Bołoz, W. 185
Bonnes, M. 35, 166
Borkenhagen, D. 183
Borowsky, B. 68, 166
Bourrez, A. 53, 167
Brabant, M. 44, 170
Bracken, D.W. 137, 167
Brannick, M.T. 35, 138, 167
Brcic, J. 184
Breaugh, J.A. 137, 167
Brenning, K. 171
Bruce, V. 41, 167
Bryce, C. 168
Buckley, T.C. 50, 167
Bush, T. 168
Buskist, W.F. 89, 167
Byrdoff, P. 143, 167
Byrne, E.A. 54, 180
Böttger, U. 179

C
Cable, D.M. 122, 167
Cacciabue, P.C. 116, 167
Cannon-Bowers, J.A.
Caplan, R.D. 124, 176, 182
Carless, S.A. 14, 121, 142, 167
Carreta, T. 137, 168
Carsten, O. 116, 190
Casner, S.M. 147, 168

Cassell, C. 184
Chan, D. 138, 181, 183
Chapanis, A. 22, 26, 34, 90, 91, 168
Chekalina, A.I. 175
Cheveigne, A. de. 38, 168
Chirkowska-Smolak, T. 138, 166
Chmiel, N. 14, 168
Christensen, J.M. 47, 168
Cieślak, R. 123, 189
Clamann, M. 58, 168
Cockell, C.S. 48, 168
Cohen, S. 38, 168
Coleine, C. 179
Collet, C. 165
Colquhoun, W.P. 52, 168
Condit, D. 53, 174
Conrad, B. 189
Convertino, V.A. 48, 168
Cook, M.J., 56, 169
Cooper, C.L. 129, 157, 170, 180
Corlett, E.N. 129, 178, 189
Covelli, V. 169
Crawford, D. 35, 169
Cunningham, D. 35, 173
Cwietkow, A. 191
Ćwiklicki, M. 169
Cymerman, A. 172
Cyr, L.A. 123, 189
Czekaj, J. 169

D
Dadachova, E. 179
Dahleberg, A. 169
Danielsson, M. 53, 184
Davis, S.E. 98, 167
Dawson, J.D. 165
De Jonge, J. 129, 188
de la Torre Noetzel, R. 48, 169
de Vera, J-P. 48, 169, 179
De Voge, J.M. 119, 169
De Voogt, A. 119, 169
Dekker, S.W. 18, 119, 166, 169

DeNisi, A.S. 137, 178
Deo, N. 166
Derlicka, M. 123, 189
Deschênes, M. 183
Desmond, P.A. 129, 171
Deutsch, S. 175
Di Majo, V. 42, 169
Direito, S. 168
Dittmar, A. 165
Dobrowolski, A. 186
Domański, H. 138, 169
Draghici, A.M. 100, 173
Draper, S.W. 56, 179
Drozdowski, R. 46, 191
Dudek, B. 131, 169
Duffey, R.B. 136, 169
Duncan, D.C. 14, 176
Dunnette, M.D. 13, 122, 169, 187
Dziuba, Ł. 182

E
Edwards, E. 125, 135, 170
Edwards, J. R. 14, 121, 122, 167, 170
Ek, Å. 135, 170
Ellifritt, T.S. 16, 22, 103, 147, 171
Ely, J.H. 31, 170
Emmerson, P. 179
Enander, A. 37, 170
Esser, P. 189

F
Fąfrowicz, M. 129, 170
Filding, N. 140, 170
Filipovic, N. 181
Fink, G. 165–168, 179, 181, 185, 188
Fischer B. 76, 166, 170
Folding, J. 140, 170
Fowler, B. 44, 45, 170, 172, 176
Fox-Powell, M. 168
Franus, E. 127, 131, 170
Friedman, G. 171

G
Gąsik, M. 180
Gaździński, S.P. 47, 171, 174
Georgeson, M.A. 41, 167
Goemaere, S. 48, 171
Göran, K. 53, 187
Gorski. A. 173
Gradwell, D. 171
Green, P.R. 41, 167
Green, R.G. 15, 22, 81, 103, 147, 171
Green, R.L. 81, 171
Gregore, A.K. 171
Guadanga, L.M. 175
Guardiera, S. 171
Guilford, J.P. 109, 137, 171
Gushchin, V.I. 175
Gushin, V. 184
Guttmann, G. 166

H
Haakonson, N.H. 134, 135, 171
Haas, M.W. 57, 189
Haddon, W. 126, 171
Hagen, F. 25, 187
Hambrick, D.Z. 177
Hamery, O. 178
Hancock, P.A. 129, 171, 172, 178, 188
Hannum, K.M 141, 171
Harris, D. 13, 18, 119, 167, 169, 171, 172, 176, 178–180, 182–184, 188
Harris, M.G. 133, 181
Harrison, J.P. 168
Hart, S.G. 129, 172
Hartong, D.T. 149, 172
Haslam, S.A. 14, 172
Hawkins, F.H. 54, 172
Hebb, D.O. 130, 172
Heinrich, H.W. 125, 172
Helmut, L. 17, 24, 190
Hengy, S. 178
Hermaszewski, M. 48, 172

Heyer, M. 58, 172
Hinson, T.A. 165
Hobbs, A. 137, 172
Holland, J.L. 14, 172
Hollnagel, E. 18, 55, 161, 172
Horne, J.A. 52, 172, 186
Hough, L.M. 13, 122, 169, 187
Houston, C.S. 45, 172
Howard, J.P. 35, 92, 93, 161, 173
Hughes, T. 179
Hull, J.G. 100, 173
Hunt, B.I. 54, 174

I
Ilyin, L. 173
Isaac, A.R. 54, 173
Ivanov, V. 42, 173

J
Jacob, P. 63, 173
Jacob, R.J.K. 74, 173
James, M. 171
Jamroz, K. 125, 173
Jamson, H. 116, 190
Janewicz, M. 174
Jasiński, T.L. 47, 120, 173, 187
Jastrzebowski, W. 16, 165
Jeannerod, M. 63, 173
Jederberg, W.W. 53, 173
Jędrys, R. 175
Jensen, R.S. 81, 173, 182
Jethon, Z. 7, 15, 21, 173
Jezierski, M. 175, 180
Johnson, B.L. 49, 173
Johnson, E.C. 121, 124, 175
Johnson, Ph.J. 184
Jones, D.R. 47, 189
Jones, L.V. 19, 137, 173
Jorritsma, F.F. 172

K
Kaber, D.B. 58, 130, 173

Kamiński, L. 166
Kane, D. 152, 174
Kanki, B.G. 137, 172
Kapuścińska, Z. 166, 168, 178, 179, 188, 191
Karn, K.S. 74, 173
Karp, M.R. 53, 174
Kearns, E. 176
Kinsbourne, M. 65, 67, 174
Klein, K.E. 53, 174, 189
Kłossowski, M. 176
Kobos, Z. 85, 109, 175, 175, 185, 186
Kocian, D.F. 57, 174
Konorski, J. 72, 86, 98, 174
Kooijman, A.C. 172
Koonce, J.M. 116, 174
Kopania, J. 158, 174
Kopczynski, K. 153, 174
Koradecka, D. 122, 174
Kotarbiński, T. 18, 174
Kowalczuk, K. 47, 174, 188
Kowalewska, S. 139, 175
Kowalski, W. 37, 111, 175, 176, 178
Kramer, M.F. 176
Krawczyk, P. 143, 175
Kristof-Brown, A.L. 121, 123, 124, 139, 175, 176
Kroemer, K.H. 92, 175
Kubiczkowa, J. 71, 175
Kuliński, J. 64, 175, 180
Kulwicki, P.V. 56, 175
Kuzak, W. 68, 115, 177
Kuznetsova, P.G. 48, 175
Kwarecki, K. 48, 53, 176
Kwiatkowski, R. 14, 176

L
Lammer, H. 168
Landenmark, H. 168
Landsbergis, P.A. 129, 188
Łaszczyńska, J. 110, 176
Latek-Najder, M. 175

Lauri, M.A. 119, 176
Lauver, K.J. 139, 176
Lefevre, R. 166
Leggatt, A.P. 179
Lempereur, I. 119, 176
Leodolter, M. 166
Leodolter, U. 166
Leuba, C. 130, 176
Levine, E.L. 26, 35, 138, 167, 176
Lewin, K. 190
Lewkowicz, R. 143, 176, 188
Li, W.C. 119, 176
Lic, T. 191
Lim, D.V. 176
Lindbeck, G. 50, 165
Lis, S.R. 82, 84, 183
Locke, E.A. 13, 176
Łomow, B.F. 95, 105, 127, 128, 177
Lopez de Alda, M.J. 150, 181
Lück, H.E. 190
Lyon, J. 183
Lyshevsky, S.E. 145, 177

M
Maciejczyk, J. 17, 68, 108, 115, 143, 177
Maijman, T.F. 130, 183
Majzel, M. 191
Makowska, Z. 169
Malawski, M. 66, 67, 72, 161, 184
Manek, A. 166
Mann, J. 166
Marcinkowski, P. 175
Marco, M.P. 150, 181
Marek, T. 129, 170
Maric, J. 181
Martin II, L.B. 36, 179
Martin, C.D. 165
Martin, W.L. (92) 58, 177
Martin-Torres, J. 168
Marty, C. 178
Mathews, G. 52, 177

Mats, G. 53, 187
McCabe, D.P. 74, 177
McCloskey, K. 188
McCormick, E.J. 33, 34, 177
McDaniel, J.W. 56, 175
McDaniel, M.A. 177
Meadows, S.K. 175
Meglino, B.M. 137, 178
Meijlaers, M. 187
Melis-Dankers, B.J.M. 172
Menon, P.E. 176
Meshkati, N. 130, 172, 178
Migdał, K. 103, 178
Mikuliszyn, R. 46, 178
Miller, J.O. 67, 72, 178
Minors, D.S. 50, 181, 189
Missiuro, W. 17, 178
Mlynczak, J. 174
Molińska, M. 44, 112, 178
Moray, N. 85, 178
Morgeson, F.P. 35, 138, 167
Mouloua, M. 188
Muir, H. 171
Mulder, M. 58, 178, 188
Mumford, M.D. 184
Münsterberg, H. 23–25, 34, 178

N
Narayanon, L. 176
Naumenko, R. 173
Navon, D. 67, 72, 178
Naz, P. 149, 178
Nazarow, A. 33, 90, 161, 178, 191
Nebylitsyn, W.D. 179
Nelson, R.J. 36, 179
Neve, J.J. 172
Nibbelke, R.J. 18, 179
Nicholson, N. 168
Nieznański, M. 77, 179
Noack, L. 168
Noppe, A. 171
Norman, D.A. 56, 179

Nowicki, G. 130, 187
Nullmeyer, R. 53, 174
Nyga, P. 153, 174

O
O'Hare, D. 127, 179
O'Malley-James, J. 168
Obidziński, M. 77, 179
Oborne, D.J. 39, 179
Ogilvie, J.R. 116, 179
Okóń, J. 33, 166, 168, 178, 179, 188, 191
Ones, D.S. 165
Onofri, S. 179
Orlansky, J. 31, 170
Oron-Gilad, T.
Orr, C.E. 68, 166
Osiński, W. 85, 179, 181
Östberg, O. 52, 172, 186
Owens, W.A. 184

P
Paas F. 118, 192
Pacelli, C. 48, 179
Palonek, H. 47, 174
Paluszkiewicz, L. 31, 33, 179
Parasuraman, R. 54, 180
Park, S. 125, 180
Patrick, J. 97, 180
Payler, S.J. 168
Pazzaglia, S. 169
Pearson, R.A. 131, 180
Perry, C.M. 173
Pietraszkiewicz, H. 166
Pińkowski, J. 85, 185
Piveteau, L.D. 152, 180
Płatonow, K.K. 105, 127, 177
Ployhart, R.E. 142, 182
Pope, A.L. 130, 180, 184
Porlier, G. 45, 170
Posner, M.I. 72, 75–78, 161, 180
Poulton, E.C. 38, 40, 41, 43, 180

Prilic, H. 44, 180
Proctor, L.J. 56, 169
Prost, M. 41, 64, 148, 175, 180
Puchalska, L. 47, 174
Pylyshyn, Z. 39, 180

Q
Qua, X. 191

R
Rabbow, E. 179
Raczek, J. 86, 180
Radakovic, S.S. 37, 111, 181
Radjen, S. 181
Rahimi, M. 130, 178
Ratajczak, Z. 127, 131, 181
Ravlin, E.C. 137, 178
Reason, J. 126, 132, 133, 181
Rebessi, S. 169
Rechnitzer, P. 35, 173
Redfern, P.H. 181
Ree, M.J. 142, 168
Reeve, Ch.L. 20, 181
Reeves, J. 172
Reinhart, R.O. 39, 43, 111, 112, 181
Reykowski, J. 127, 181
Reynolds, S. 121, 181
Ridder, L.D. 187
Rigner, J. 18, 169
Rizzo, M. 165
Robertson, I.T. 170
Rock, P.B. 133, 181
Rodriguez-Mozas, S. 150, 181
Roediger, H.L. 177
Roessingh, J.M. 116, 181
Rogelber, S.G. 29, 181
Rollag, K. 139, 181
Romero, L.M. 50, 181
Roth, J. 36, 37, 182
Różanowski, K. 151, 182
Ruddle, R.A. 18, 55, 182
Ruitenberg, B. 54, 173

Rushby, A. 168
Russo, J.E. 76, 161, 182
Ryan, A.M. 142, 182
Ryś, M. 185

S
Sadowski, B. 40, 182
Salas, E. 124, 182
Salden, R.J.C. 119, 182
Samel, A. 53, 182
Samuels, T. 168
Saran, A. 169
Sarapata, A. 138, 175, 182, 187
Sargent, J.D. 100, 173
Saull, J.W. 136, 169
Sawinski, Z. 138, 169
Scanlon, M.V. 149, 182
Schaumburg, H. 49, 183
Schellekens, J.M. 130, 183
Scherbaum, C.A. 138, 140, 183
Schlosberg, H. 27, 31, 32, 71, 74, 86, 88, 89, 190
Schmitt, N. 138, 183
Schneider, S. 171
Schuhfried, G. 143, 166, 183
Schultz, D. 14, 97, 183
Schultz, S.E. 14, 97, 183
Schwendner, P. 168
Scialfa, C.T. 114, 183
Secchiaroli, G. 35, 166
Sechmenov, I.M. 183
Segall, N. 173
Selbmann, L. 179
Self, H.S. 15, 22, 103, 147, 171
Shannon, C.E. 23, 89, 183
Shappell, S.A. 137, 189
Sheik-Nainar, M.A. 173
Shelhamer, M. 49, 185
Sherman, P.J. 124, 183
Shi, Q. 165
Shimmin, S. 14, 176
Shved, D.M. 175

Sienkiewicz, Z. 166
Simons, H.W. 13, 183
Simpson, J.M. 176
Sinangil, H.K. 165
Singleton, W.T. 131, 135, 183
Skibniewski, F.W. 68, 115, 151, 177, 182
Słomczyński, K.M. 138, 169
Smallwood, T. 97, 183
Snook, S.H. 175
Sobotnicki, A. 174
Spacapan, S. 38, 168
Spector, P.E. 176
Spencer, P. S. 48, 183
Spohn, T. 48, 169
Stankovic, N. 181
Stark, L.L. 82, 84, 183
Starke, M. 137, 167
Starr, A. 179
Stasiak, K. 64, 175, 183
Staveland, L.E. 129, 172
Stefanova, E. 181
Stephens, Ch.L. 130, 184
Still, K. 53, 173
Stokes, G.S. 140, 184
Strelau, J. 127, 184
Strüder, K. 171
Stubbs, J. 53, 184
Suedfeld, P. 48, 184
Sundstrom, E. 13, 14, 18, 184
Surbatovic, M. 181
Sutton, J.R. 172
Świątek, H. 103, 115, 184
Święcicki, W. 176
Symon, G. 16, 184
Szajner, R. 113, 190
Szczechura, J. 66, 67, 72, 73, 77, 79, 81, 83, 85, 88, 103, 115, 136, 161, 184–186
Szmigielski, S. 49, 185
Szocik, K. 48, 185
Szumilewicz, J. 166

T
Talbot, J.M. 47, 168
Taoa, D. 191
Tarnowski, A. 77, 79, 99, 130, 166, 185, 186
Taylor, F.W. 25, 185
Taylor, M. 37, 45, 185
Terelak, J.F. 17, 18, 46–49, 73, 77, 79, 81, 83, 85, 88, 94, 99, 101, 103, 105, 109, 114, 115, 120, 123, 136, 140, 143, 170, 176–178, 184–187
Thierens, H. 42, 187
Thomas Aquinas, St. 21, 165
Thomson, R.M. 31, 170
Thorndike, R.L. 25, 187
Tichon, J. 114, 187
Tinsley, H.E.A. 122, 187
Tomaszewski, T. 26, 98, 106, 187
Torbjörn, A. 53, 187
Trapeznikov, W.A. 165
Triandis, H.C. 13, 187
Truszczyński, O. 46, 47, 109, 114, 124, 130, 186–188
Tsang, M.S. 97, 132, 168, 180, 183, 188
Tubiana, M. 42, 151, 188
Tukov, A. 173
Turski, B. 39, 188

U
Uc, E.Y. 165
Uszakowa, M. 87, 188

V
Værnes, R.J. 45, 112, 188
Van der Pal, J. 119, 182
Van der Vaart, H.J.C. 58, 178
Van Doorn, R.A. 119, 169
Van Paassen, M.M. 58, 188
Vansteenkiste, M. 171
Vegchel, N. 129, 188
Vermeulen, A.C.J. 171

Vernet-Maury, E. 165
Vidulich, M.A. 97, 132, 168, 180, 183
Vincenzi, D.A. 116, 188
Vinokhodova, A.G. 175
Viswesvaran, C. 165
Von Gierke, H.E. 39, 46, 188
Vral, A. 187

W
Wadsworth, J. 168
Walichnowski, W. 112, 188
Wallace, P. 157, 189
Wannok, S. 145, 189
Waterhouse, J.M. 50, 181, 189
Weaver, W. 23, 183
Weber H. 29, 76, 170
Wegmann, H.M. 53, 182, 189
Wells, M.J. 57, 189
Westerman, J.W. 123, 189
Whinnery, J.E. 47, 189
Wickens, C.D. 65, 189
Widerszal-Bazyl, M. 99, 123, 189
Wiegmann, D.A. 126, 189
Wilde, G.J.S. 126, 189
Williams, H. 127, 183, 189
Wilson, J.R. 129, 178, 189
Wilson, K. 56, 169

Winter, B.J. 113, 189
Wise, J.A. 188
Wochyński, Z. 175
Wojtkowiak, M. 47, 112, 113, 188, 190
Wolff, S. 41, 190
Woodworth, R.S. 27, 31, 32, 71, 74, 86, 88, 89, 190
Wyleżoł, M. 174

X
Xuec, H. 191

Z
Żakowska, L. 116, 190
Zalewska, A. M. 140, 190
Zawadzki, B. 17, 142, 178, 190
Żebrowski, M. 46, 178
Zeidler, W. 16, 17, 24, 190
Zhanga, X. 52, 191
Zickar, M.J. 15, 191
Zieliński, P. 46, 191
Zimmerman, R.D. 121, 124, 175
Zinczenko, W. 139, 191
Zorzano, M.P. 168
Zucconi, L. 179
Zużewicz, K. 176

Studies in Social Sciences, Philosophy and History of Ideas

Edited by Bogusław Paź

Vol. 1 Józef Niżnik: Twentieth Century Wars in European Memory. 2013.
Vol. 2 Szymon Wróbel: Deferring the Self. 2013.
Vol. 3 Cain Elliott: Fire Backstage. Philip Rieff and the Monastery of Culture. 2013.
Vol. 4 Seweryn Blandzi: Platon und das Problem der Letztbegründung der Metaphysik. Eine historische Einführung. 2014.
Vol. 5 Maria Gołębiewska / Andrzej Leder/Paul Zawadzki (éds.): L'homme démocratique. Perspectives de recherche. 2014.
Vol. 6 Zeynep Talay-Turner: Philosophy, Literature, and the Dissolution of the Subject. Nietzsche, Musil, Atay. 2014.
Vol. 7 Saidbek Goziev: Mahalla – Traditional Institution in Tajikistan and Civil Society in the West. 2015.
Vol. 8 Andrzej Rychard / Gabriel Motzkin (eds.): The Legacy of Polish Solidarity. Social Activism, Regime Collapse, and the Building of a New Society. 2015.
Vol. 9 Wojciech Klimczyk / Agata Świerzowska (eds.): Music and Genocide. 2015.
Vol. 10 Paweł B. Sztabiński / Henryk Domański / Franciszek Sztabiński (eds.): Hopes and Anxieties in Europe. Six Waves of the European Social Survey. 2015.
Vol. 11 Gavin Rae: Privatising Capital. The Commodification of Poland´s Welfare State. 2015.
Vol. 12 Adriana Mica / Jan Winczorek / Rafał Wiśniewski (eds.): Sociologies of Formality and Informality. 2015.
Vol. 13 Henryk Domański: The Polish Middle Class. Translated by Patrycja Poniatowska. 2015.
Vol. 14 Henryk Domański: Prestige. Translated by Patrycja Poniatowska. 2015.
Vol. 15 Cezary Wodziński: Heidegger and the Problem of Evil. Translated into English by Agata Bielik-Robson and Patrick Trompiz. 2016.
Vol. 16 Maria Gołębiewska (ed.): Cultural Normativity. Between Philosophical Apriority and Social Practices. 2017.
Vol. 17 Anita Williams: Psychology and Formalisation. Phenomenology, Ethnomethodology and Statistics. 2017.
Vol. 18 Mikołaj Pawlak: Tying Micro and Macro. 2018.
Vol. 19 Franciszek Sztabiński / Henryk Domański / Paweł B. Sztabiński (eds.): New Uncertainties and Anxieties in Europe. Seven Waves of the European Social Survey. 2018.
Vol. 20 Adriana Mica / Katarzyna M. Wyrzykowska / Rafał Wiśniewski / Iwona Zielińska (eds.): Sociology of the Invisible Hand. 2018.
Vol. 21 Jan Felicjan Terelak: Psychology of the Operator of Technical Devices. 2020
Vol. 22 Dorota Maria Leszczyna: Del idealismo al realismo crítico. La política como realización en José Ortega y Gasset. 2020
Vol. 23 Adam Olech: The Semantic Theory of Knowledge. 2020
Vol. 24 Zbigniew Drozdowicz: La république des savants. Sans révérence. Traduit du polonais par Catherine Popczyk. 2020

Vol. 25 Andrzej Waśkiewicz: The Idea of Political Representation and Its Paradoxes. Translated from Polish by Agnieszka Waśkiewicz and Marilyn Burton. 2020

www.peterlang.com

www.ingramcontent.com/pod-product-compliance
Ingram Content Group UK Ltd.
Pitfield, Milton Keynes, MK11 3LW, UK
UKHW041902230426

12049UKWH00002B/14